Finding the First Europeans: Archeology on the Lower Alabama River

(The 2015–2016 Seasons)

by: Caleb Curren

Contact Archeology Inc.
2016

Contents

Contents
(cont.)

Illustrations

Illustrations
(cont.)

Illustrations
(cont.)

Abstract

The archeological site of the largest battle ever fought between Europeans and Native peoples in North America has not been found, neither has the site of the first long-term European colony in the interior of the current United States. They are both located in the present state of Alabama.

Both sites are benchmarks in American Archeology. The sites are capable of providing us with unprecedented data relative to the Natives and Spaniards of that time period. Both sites are located in Southwest Alabama. Contact Archeology Inc., a nonprofit research and education group of specialists, is searching for both of those premier sites.

The writings of the Spaniards told of a small Native town named "Mabila." The documents state that a great battle between thousands of Native peoples and some six hundred Spaniards was fought at that Native town on October 18th of 1540. Many people were killed. The Spanish survivors were all wounded. The battle changed the lives of the Spaniards and the Natives as well as their descendants.

Some twenty years later, Spaniards from a second expedition wrote about another Native town. It was called Nanipacana. The Spaniards came to the Southwest Alabama Native town from their colony on Pensacola Bay in October of 1559. The Natives abandoned the town to the Spaniards who renamed it the "Settlement of the Sacred Cross."

The impetus of both the 16th-Century expeditions led by Soto and Luna originated with the Spanish Royals in Spain. Those Royals wanted to explore and lay claim to this land of the current southeastern United States. They also desired to procure any riches in the region, and establish settlements to protect their fleets of treasure ships sailing from the Americas to homeland Spanish ports.

The Spanish were expanding their territories. The Native peoples were doing the same as they had been for centuries. The clash of the two cultures led to some of the most dramatic human encounters in the history of the Western Hemisphere. Ironically, the writings of the Spanish Conquistadors are the best record we have concerning the Native peoples they were sent to conquer.

When the sites of Mabila and the Settlement of the Sacred Cross are found, the result will be a significant connection between the Spanish entradas and the lifeways of the Native peoples. This following document is a step in that direction.

Acknowledgements

The following individuals and organizations have provided one or more of the following contributions to this unique Southwest Alabama early Spanish Contact project: Sponsorship funds, property access permission, volunteer time, scientific data, images, lab space and/or equipment, community contacts … and perhaps most important of all, the vital encouragement in times when the bugs and heat are oppressive or we're waiting on landowner access permission or the midsummer storms hit or the roads are impassible or the river is high or the boat is leaking or the trailer has a flat, or all at the same time. Our thanks to all these people.

Sponsorships:

Nicholas H. Holmes Jr.
Ann Bedsole Holmes
Nicholas Holmes III
Edmon H. McKinley
M.W. Smith Jr. Foundation
Thomas L. Turner Charitable Trust
Mobile Historic Development Commission

Property Access Permission:

Slaughter Family
Parham Family
Alabama Trust Fund
Booth Family
Meador Family
Broughton Family

Space and/or Equipment:

Tower East Group
Frontier Motors
Pensacola Hardware
Great Southern Restaurants
Weatherford's Outfitters
East Hill Hardware
Jolly Sailing Charters
Walmart

Images:

University of South Alabama
University of Alabama
Alabama Historical Commission
Google Public Domain

Acknowledgements
(cont.)

Community Liaisons:

Terry Berling
Louis Finlay
Charlie Clark
George Alford

Field Team Members:

Joni Eddins
Mark Martine
Eugene Wilson
Jeffrey Keiek
Shawn Enfinger
Aiden Enfinger
Andrew Gehlken
Jeaney Patti
Mary Snyder
Dana Arduini
Alex Arduini

Scientific Data Contributors:

Ned Jenkins, Fort Toulouse / Jackson Park
Alabama Historical Commission
Charlie Clark, U.S. Department of Agriculture
John Powell, Historic St. Augustine
Alabama State Site Files, University of Alabama
U.S. National Museum, Smithsonian Institution
Chester DePratter, S.C. Institute of Anthropology / Archaeology
Keith Little, Tennessee Valley Archaeological Research
David Hurst Thomas, American Museum of Natural History

Contact Archeology Inc.

Contact Archeology Inc. was formed to bring together a core group of specialist consultants in order to pool their talents toward the discovery of these "First Contact" sites in Alabama and Florida. The principles in the group are:

Caleb Curren: President

David Dodson: Historian

Michael O'Donovan: Communications Director

Nicholas H. Holmes, Jr.: Architect

Eugene M. Wilson: Geographer

William Overman: Treasurer

Jerry Gill: Copy Editor

Joni Eddins: Lab Supervisor

Mark Martine: Field Supervisor

Jeffrey Keiek: Field Supervisor

Cindy Justice: Logistics Coordinator

Roxanne Rachel Lavelle: Website

Terry Berling: Florida Community Liaison

Louis Finlay: Alabama Community Liaison

Charlie Clark: Alabama Community Liaison

George Alford: Alabama Community Liaison

Jim Bielinski: Boat Captain

Contact Archeology Inc.
P.O. Box 30506, Pensacola, Florida 32503
website (archeologyink.com)

A Federally recognized 501 (c) (3) corporation.
DLN: 17053084358013
Public Charity Status: 170 (b) (1) (A) (vi)

Introduction

The European Contact Period of the mid-1500s was one of the most potent times in the history of the current United States. European adventurers, treasure seekers, and colonists penetrated the interior of the region with unprecedented vigor.

The numbers of Europeans were small but larger than the Native peoples had ever seen. It was the first time in over 10,000 years of cultural development in that vast region that Native peoples had seen so many foreigners for so long a period in their homeland.

Two Spanish expeditions arrived in the mid-1500s. One was led by Hernando de Soto in 1539. Another was led by Tristan de Luna in 1559. The impacts that the Native and European cultures had on one another were multiplistic and long lasting, even into the present day.

Scholars and laymen have been searching for archeological sites related to both those Spanish expeditions for over 150 years. This publication presents an overview of archeological and historical searches in South Alabama and Northwest Florida relative to the two Spanish expeditions. We begin in the present.

During the year 2015-16, Contact Archeology Inc. obtained sponsorships and grants to conduct archeological field and laboratory research focused on 16th-Century Spanish and Native Mississippian Period sites on the lower Alabama River and Northwest Florida. The research strategy involves historical documents research in conjunction with archeological field tests.

The document research revolves around historian David Dodson's fourteen years of examining maps and documents in numerous archives in Europe, the United States, Mexico, and the Caribbean. A number of early published sources are also used in the research (Priestley 1929, vols.1-2; Priestley 1936; Padilla 1596; Bourne 1904 (a-b); Varner and Varner 1951).

In addition to the archeological testing and historical document research, necessary activities include obtaining landowner permissions, revisiting previously recorded archeological sites, accessing private artifact collections, and following leads on discoveries by local people.

Previous archeological Contact Period research was conducted in the 1980s and 1990s on the lower half of the Alabama River through the auspices of several agencies (Curren 1992). The research resulted in one of the most comprehensive archeological studies ever conducted on the lower Alabama River. The 2015-16 research season utilizes much of the data collected during those earlier decades.

This current document presents an overview of the stories of the Spanish expeditions of Soto (1539) and Luna (1559), focusing on the long-sought archeological sites of Mabila and Nanipacana on the Alabama River and Santa Maria de Ochuse on Pensacola Bay.

In keeping with proper Spanish language usage, "Soto and Luna," rather than "de Soto and de Luna," is used in this document.

The sections of this document include: an introduction to the research issue, field and lab techniques, objectives of the project, archeological and historical data relative to the project, individual site synopses, an evaluation of the 2015-16 research season, and a section of references and related works for those who wish to conduct their own expanded research into the subject of the earliest European and Native contacts in the current Southeastern United States.

Be it known that this is the farthest south we have ever searched for the archeological sites of Mabila and Nanipacana. Our critics think us somewhat foolish. In support of our critics, the estimated distances given us by the Spanish chroniclers do not support our surveys this far south down the Alabama River. However, researchers have looked for Mabila and Nanipacana for over a hundred years upriver on the Alabama River drainage without success.

Our new hypothesis testing farther downriver may or may not provide the necessary archeological criteria for the validation of the locations of the battle site of Mabila and the Spanish site of the "Settlement of the Sacred Cross."

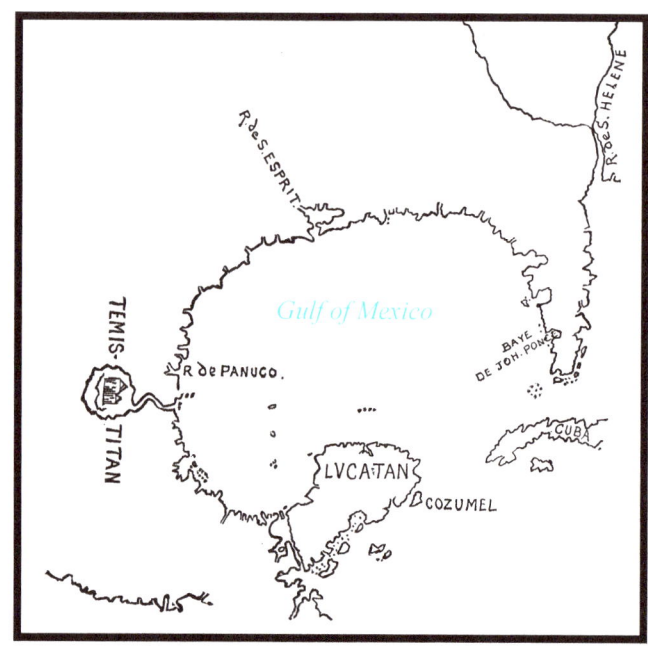

Figure 1: 16th-Century Spanish Map. Adapted from Public Domain Image.

Swanton stated emphatically that:

"Correct or Official"

One of the biggest issues for researchers delving into Contact Period studies is the temptation to prematurely proclaim specific archeological sites or specific expedition routes as fact before adequate archeological testing is completed.

In the excitement of the moment of anticipated discovery, it is very tempting to jump to conclusions and proclaim a site or an expedition route "official" before thorough archeological hypothesis testing confirms or refutes the postulates.

Personnel from the University of Alabama, the University of Georgia, the U.S. National Park Service, and the Alabama De Soto Commission made that mistake with their premature "official" marking of the Soto route through Alabama and the Southeast during the late 1980s and early 1990s.

Personnel from the University of West Florida also made the same mistake in 2015 by prematurely "officially" claiming … emphatically … both nationally and internationally, that they had located the Luna Colony at site 8Es1 on Pensacola Bay. The site may indeed, eventually, prove to be the "Correct" Luna Colony but the initial "Official" colony claim came before appropriate absolute hard data had been found.

This credibility issue was best stated by the venerated Dr. John R. Swanton, chairman of the 1935 Federal Commission, who was assigned by President Franklin D. Roosevelt the daunting task of tracing the route of the Hernando de Soto army through the Southeast. Dr.

*... **Nothing is made correct because it is called "Official."***

It becomes recognized as official by the virtue by which it is worked out, and the evidence you can focus upon it. If the work is recognized and carefully done, if every possible line of evidence is brought to bear upon it and it is permanently accepted by historians, students, etc.,
... then it is accepted and becomes official.

Dr. John R. Swanton, PhD.
Harvard University,
Chairman of the United States De Soto Commission
Appointed to the Commission by
President Franklin D. Roosevelt, 1935

Figure 2: Dr. John R. Swanton.
Public domain image.

(Quote from the official minutes of the Commission, May 4-5, 1936.) (source: *The Foreword*, by William C. Sturtevant, in *Classics of Smithsonian Institution, Final Report of the United States De Soto Expedition Commission by John R. Swanton*, Introduction by Jeffrey P. Brain, Smithsonian Institution Press, Washington, DC, 1985)

Field Techniques

Following Dr. Swanton's lead, it is imperative that empirical field techniques are utilized. These techniques are multiplistic in nature to bring "every possible line of evidence" to bear on the objective of locating Spanish sites and applying that knowledge to the early historic period of the Southeast.

Archeological field research cannot be conducted if you can't get permission to get on the land. Hence, a significant portion of the 2015 season was spent making connections with landowners in the lower Alabama River drainage. This entails engaging local people. It is necessary to convey to landowners the exact mission of Contact Archeology Inc. We do not want to take their land. We do not want to take valuables from their land.

We simply want to gain access to their properties to further our studies of the early history of the region. It takes time and patience to gain the trust of the landowners.

Once land access is granted, we walk the properties, examine river bank profiles, dig shovel tests, take photographs, record field notes, and bring the data back to the laboratory for interpretation. We live in a world of marvelous technology and we use it at Contact Archeology Inc., yet still, dirt must be moved to find the Spaniards.

Figure 3: Field Photos.
Contact Archeology Inc. Files.

4

Lab Techniques

The laboratory facilities of Contact Archeology Inc. are housed in the historic Sacred Heart Hospital building in Pensacola, Florida. The building is listed on the National Register of Historic Places. It is a Gothic Revival architectural rare gem and was built by some one hundred men in approximately one year. Some of the men were master stonemasons from Europe. The building is unique.

Our laboratory procedures follow standard, accepted practices for professional archeological studies. Provenienced artifacts from field surveys and testing are returned to the lab where they were cleaned, tagged and bagged, identified, entered into computer files, photographed, and placed in accessible containers or put on display.

Figure 4: Laboratory Photos.
Contact Archeology Inc. Files.

Objectives of the Research Project

Contact Archeology Inc. is conducting archeological and historical research to locate the site of the 1540 Spanish and Native battle site at the town of Mabila. We also plan to locate the Native town of Nanipacana, site of the 1559 Spanish Luna Settlement of the Sacred Cross. The criteria for each site are listed below. The criteria are based on 16th-Century Spanish writings and archeological projections.

Mabila Battle Site, Archeological Criteria

1. Archeological features indicating a small Native town.
2. Postholes and daub indicating a burned palisade surrounding the town.
3. A pond of fresh water near the site.
4. One gate on the east end, another on the west.
5. A featureless vacant area (plaza) in the center of the town site.
6. Postholes outlining large structures within the palisade.
7. Fire hearths containing 16th-Century Spanish artifacts.
8. Spanish burials. At least twenty Spaniards died there.
9. Faunal remains of horses and pigs. Most likely teeth.
10. 16th-Century Spanish artifacts in the town and around it (the battle).

Settlement of the Sacred Cross, Archeological Criteria

1. Archeological features indicating a large Native town.
2. Conspicuous hill adjacent to the site.
3. Spanish burials. Unknown number of Spanish colonists died there.
4. Fire hearths with Spanish artifacts in and around them.
5. Site adjacent to or near the east side of the Alabama River.
6. Spanish artifacts, likely small, scattered across the site.

The following sections of this document address the histories of both expeditions.

Two Early European Contact Expeditions

(Bourne 1904 a-b; Varner and Varner 1951; Padilla 1596; Priestly 1928, 1936; personnel communication, David Dodson 2014-2016)

When people are asked today about the first beginnings of the current United States, most think of English Jamestown in Virginia or Plymouth and the Mayflower in New England during the 1600s, or Spanish St. Augustine in the 1500s. These misconceptions are brought about from well-intended textbooks of our youth.

In reality, the origin of the current United States began over 1,000 miles south of Jamestown and Plymouth.

Two major Spanish expeditions were launched into the current Southeastern United States during the mid-1500s.

In May 1539, a 600-man army led by Hernando de Soto plunged into the Southeast from a bay along the central Florida coast, beginning a four-year odyssey searching for riches of gold, silver, and precious stones.

In August 1559, a 1500-person expedition led by Tristan de Luna landed on Pensacola Bay, ordered by the Spanish King to establish the first long-term European colony in this country.

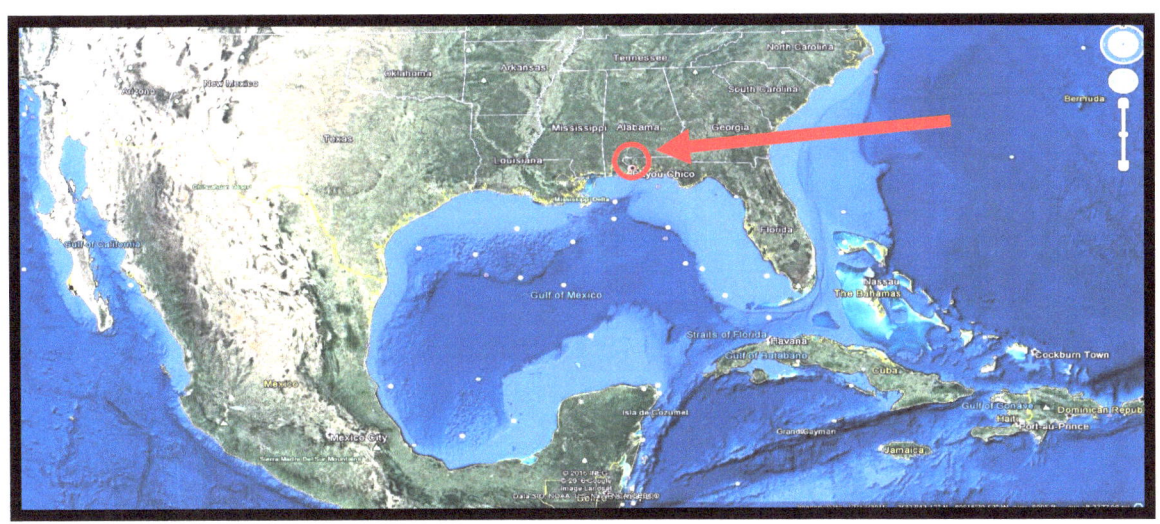

Figure 5: General Location of Project.

The researchers of Contact Archeology Inc. propose that both those Spanish expeditions passed through portions of Southwest Alabama and Northwest Florida. Consequently, the nonprofit research and education group has investigated over 100 sites in those areas.

The research effort is a collaboration of archeologists, historians, translators, foundations, individual sponsors, landowners, and volunteers.

To put the archeology in context with the history, it is important to understand that the Spanish Empire was expanding its domain and wealth during the 1500s in the Southeast. The Native Chiefdoms were doing the same in their homeland. With such territorial expansion by both groups, the clash between the two continents was inevitable. Consequently, a series of remarkable events unfolded.

The Spanish armies did not find peaceful "children of the forests" in this new country. The Native peoples were fierce warriors. Unarmed Spaniards carrying white flags of peace would have been slaughtered by that warrior society.

Three of the most important archeological sites of the earliest European Contact Period are located somewhere in Southwest Alabama and Northwest Florida. Remarkably, these pivotal sites have not been found.

First, the largest battle ever fought between Europeans and Natives in the current United States erupted in 1540 in Southwest Alabama at a Native town named Mabila.

Second, the first long-lived European Colony in this country was established on Pensacola Bay, Florida, in 1559.

Third, the first long-lived European settlement in the interior of the current United States was established in Southwest Alabama in 1559.

Researchers have tried for over a century to locate these archeological sites to no avail. The following pages detail the research design planned by Contact Archeology Inc. to locate the three sites. A summary of each of these 16th-Century Spanish expeditions is also provided.

Figure 6: Spanish Helmet,
Public Domain Image.

The 1539 Soto Expedition

(Bourne 1904 a-b; Varner and Varner 1951)

The Hernando de Soto army of 1539 was composed of 9 ships, approximately 600 people, 243 horses, and a herd of hundreds of pigs. Only about half the army and 25 horses survived the expedition. The pigs survived and multiplied. Their descendants still root up farmers' fields and hunters' game plots to this day.

The majority of the Soto army were men from Spain. Others came from Portugal, France, Italy, Greece, Cuba, and Africa. At least two women were on the expedition. One of the women was burned to death during a battle with Natives. Her name was Francisca. Another woman managed to survive the entire expedition from Florida to Mexico. She was a serving woman to the Spanish upper class. She returned to Cuba from Mexico after the expedition. She then disappeared from the historical record, taking her incredible story with her.

Figure 7: Historical Woman. Public Domain Image.

The Soto expedition was extremely well funded. Hundreds of thousands, if not millions of dollars in today's money, launched the endeavor. The professions of the participants included mounted knights, foot soldiers, sailors, shoemakers, sword-cutlers, tailors, carpenters, notaries, trumpeters, ship caulkers, priests, and more.

The fortunes of the expedition took a dramatic turn on a 1540 autumn morning at a small Native town in current southwest Alabama.

The 1540 Battle Site of Mabila

(Bourne 1904 a-b; Varner and Varner 1951)

On the morning of October 18th, 1540, the 600-man Spanish army led by Hernando de Soto entered the small, heavily fortified Native town of Mabila in Southwest Alabama. A large and powerful army of Native warriors, thousands strong, had set a trap for the Spaniards at the town. Shortly after the Spaniards entered the town a pitched battle erupted and lasted throughout the day.

In the end, the Spanish held the field but every man in the army was wounded. Thousands died that day … Knights of Spain along with their mounts, Native warriors, and Native women who helped lure the Spaniards into the town. Many of the women fell while wielding the weapons of their dead kinsmen.

The remnants of the Spanish army spent the night amidst smoldering ruins of the town that had once been Mabila. The soldiers of the army licked their wounds for a month afterwards.

Supply ships awaited them in Pensacola Bay not far to the south. The army never reached those supply ships.

Soto turned his army north for fear of losing his men to desertion once they saw their ticket home in the sanctuary of the ships.

Half of the army and all the horses died before they reached the safety of Mexico City some two years later. The survivors were almost unrecognizable beneath their animal skin clothes, scruffy beards, and bedraggled appearance. Once the men were recognized, they became national celebrities, eventually moving on with their lives.

The location of that huge battle at the small Native town of Mabila in Southwest Alabama is still a mystery to this day. We are looking for it.

Figure 8: Spanish Conquistador Army, Public Domain Image, National Park Service De Soto National Memorial.

The 1559 Luna Colony on Pensacola Bay

(Padilla 1596; Priestly 1928, 1936; personal communication, David Dodson 2014-2016)

Two decades after the Mabila Battle, the King of Spain ordered another expedition to return to the Southeast.

The King left the logistics of the expedition to his Viceroy of Mexico. The Viceroy chose a friend of his to lead the expedition, Don Tristan de Luna. His friend was a veteran of the Coronado expedition into the southwestern region of the current United States twenty years earlier. He seemed qualified for the new expedition.

Luna launched his expedition into the Southeast in 1559. It consisted of 11 ships, some 1,500 people, and 240 horses. The fleet ended up anchoring in Pensacola Bay, the same bay that saw ships anchored from the Soto Expedition.

Approximately 500 of the Luna Colony men were soldiers, cavalrymen, crossbowmen, shield bearers, and even Aztec warriors from Mexico. The rest were colonists, including men, women, children, Africans, priests, and artisans to build the colony.

Some married women with children had come with their husbands as colonists in hopes of a better life. They ended up keeping their loved ones alive by collecting roots and berries.

The settlers' mission, commanded by the King, was to establish a colony on the bay. The presence of that colony was a link in the chain that was designed to help protect the Spanish treasure fleets sailing to Spain from the Americas.

Despite a hurricane, hunger, Native attacks, disease, and internal strife, the Luna Colony lasted for over two years, the longest time ever for a European Colony up to that time.

Spanish artifacts from the 1500s have been found at six different land sites along the Pensacola Bay shore. Two shipwrecks from the Luna Expedition have been discovered in the bay, grounded and sunk in a storm on a shallow sand shelf. The anchorage of the Luna fleet has not been found.

An archeological site (8Es1) on Pensacola Bay has officially been claimed by the University of West Florida (UWF) as the Luna Colony site. However, all of the archeological criteria for such a claim have not been demonstrated.

The necessary Luna Colony criteria are:

1. **Spanish burials** reflecting the people who died there.
2. **Fire hearths** with Spanish artifacts in and around them.
3. **Spanish artifacts,** likely small, scattered across the site.
4. **Spanish refuse pits** containing predominantly Spanish artifacts.
5. **Spanish structures** defined by posthole patterns.

Figure 9: UWF Excavations at 8Es1, the alleged Luna Colony Site. Contact Archeology Files.

As of near the end of August 2016, 1000+ shovel tests having been dug at the site by UWF (personal communication, John Worth 8/22/16). Some of the shovel tests have been expanded into larger excavation units. Residential construction trenches have also been monitored. With all that impressive amount of excavation, only one possible structure with a "floor," several postholes, and an adjacent large refuse pit have been discovered. The possible structure and refuse pit may or may not be from the Luna Colony.

Sixteenth-Century Spanish artifacts have also been found scattered over the site. The artifacts, mostly pottery sherds, have been found mixed with Native pottery sherds. The site has been known as a multicomponent Native village since the Smithsonian Institution first reported it in 1883.

Native ceramics from this Woodland and Mississippian site indicate a relatively lengthy occupation period that included the 1500s. It is likely that the Natives traded with various Spanish expeditions during portions of that period.

Two Native burial mounds at the site were reported by the Smithsonian in 1883. Typical Spanish trade goods such as glass trade beads and bells found in close proximity in an area of site 8Es1 reinforce this "trade goods in burial mounds" pattern found at numerous sites in the Southeast. The Spanish artifacts likely represent Native burial goods from one of the mounds that was likely flattened during 20th-Century construction activities.

Besides the burial mound scenario, the Native peoples certainly could have easily salvaged objects from the two Luna shipwrecks lying just offshore in 10-15 feet of water after the Spanish colonists left Pensacola Bay. The Natives could have brought the salvaged Spanish objects back to their village at 8Es1.

The Native salvage and trading could explain the mixed Native and Spanish artifacts at the site.
Thus, there are alternative explanations for the presence of 16th-Century Spanish artifacts at the 8Es1 site other than the official claim of the Luna Colony.

In summary, the University of West Florida has designated site 8Es1 as the indisputable location of the 1559 Spanish Luna Colony. That claim has been disseminated through numerous news outlets and, at least, one research paper presented at a professional conference.

In the view of Contact Archeology Inc., there is not enough archeological evidence at this time to make it "official or correct" that site 8Es1 is the Luna Colony site.

A better approach by UWF to the situation would have been to propose the hypothesis that 8Es1 might be the Luna Colony site and proceed to test the hypothesis. That, unfortunately, was not the approach taken by UWF.

Hopefully, additional excavations at the site by UWF will provide the necessary archeological criteria to legitimately validate site 8Es1 as the "official" location of the 1559 Luna Colony on Pensacola Bay.

Until then, the Luna Colony claim by the University of West Florida remains "official" but is yet to be proven "correct."

The 1559 Luna "Settlement of the Sacred Cross"

*(Padilla 1596; Priestly 1928, 1936; DeJarnette et al. 1975; Curren 2014;
personal communication, David Dodson 2014-2016)*

Exploratory groups from the Pensacola Bay Luna Colony managed to penetrate the interior as far as northeast Alabama, and establish the first long-lived European settlement in the interior of the current United States on the lower Alabama River in 1559.

A total of about 800 soldiers and colonists (men, women, and children) managed to travel by water and land from the Pensacola Bay mother colony to the lower Alabama River settlement in October of 1559.

Spanish scouts had previously found a suitable settlement site at a Native town named Nanipacana. The Natives abandoned the town when they saw the main body of the Spaniards approaching. The Spanish renamed the abandoned town the "Settlement of the Sacred Cross." They set up their camp there.

The colonists maintained the settlement for about eight months before they took to their boats and rafts at the end of June 1560 to travel down the Alabama River to today's Mobile Bay to await the arrival of supply ships from Mexico. They set up camp at the head of Mobile Bay in July 1560 and stayed for approximately one month.

That Mobile Bay Luna Colony campsite has, likely, been identified by Contact Archeology at the head of the bay. The location of the "Settlement of the Sacred Cross" on the Alabama River is still a mystery.

Figure 10: Archeological Excavations at the Probable Luna Campsite on Mobile Bay. Contact Archeology Inc. Files.

Figure 11: Woman with Child. Public Domain Image.

Assessment:
The Soto and Luna Expeditions

Was the Soto Expedition a failure? Well … yes and no. They didn't find the gold and half the army died. That would be considered a failure. However, the army managed to survive in an unexplored hostile land for some four years and brought back invaluable written descriptions of Native cultures living in the interior of the current United States. That had never been done before. That is the true treasure they brought back with them. That was not their intent but, fortunately for modern scholars, it worked out that way.

Was the Luna Expedition a failure? Well … yes and no. The colonists failed to establish a permanent colony on the Gulf Coast and the interior. However, that group of Spanish men, women, children, and Aztec warriors managed to maintain a European presence in a foreign land for over two years, longer than any settlement attempt before them. That was a victory whether they knew it or not.

There are many more archeological and historical questions than answers concerning the first European Expeditions into Southwest Alabama and Northwest Florida. The mysteries are many: Where is the 1540 Soto battle site of Mabila in Southwest Alabama? … the 1559 Luna Settlement of the Sacred Cross in Southwest Alabama? … the anchorage of the 1559 Luna fleet on Pensacola Bay? … the 1559 Luna Colony on Pensacola Bay?

Clues to these mysteries can be found in dirt-covered artifacts and dusty historic documents. Contact Archeology is brushing off the dirt and the dust.

Figure 12: Excavation Unit, Pensacola Bay.
Luna Colony Search,
Contact Archeology Inc. Files.

The Previous Searches for Mabila and Nanipacana

(Picket 1851; Ball 1879; Swanton 1939; Lankford 1977; Curren 1984, 1987, 1992, 2007; Blake 1988; Brain 1985; Hudson et al. 1985; Hudson et al. 1990; Holmes 2004; Jenkins and Paglione 2007; Trickey 1995; Holmes 1993; Knight 2009; Wilson et al. 2007)

Researchers have been hypothesizing (educated guessing) the location of the Soto expedition battle site of Mabila and the Luna settlement site for over a hundred and fifty years. Numerous locations have been suggested as the sites of Mabila and Nanipacana. The hypothesized archeological sites of the towns have ranged, geographically, from the Selma, Alabama, area to the lower Alabama River region.

In 1851, historian Albert Pickett proposed a site on the lower Alabama River in Clarke County as the Mabila battle site. Archeological testing disproved the hypothesis.

In 1879, historian T.H. Ball refuted the Pickett location and proposed another site on the lower Alabama River Clarke County as Mabila. The hypothesis has since been disproved.

In 1939, the U.S. Government published the findings of the "United States De Soto Expedition Commission." The Commission was chaired by Dr. John R. Swanton. Concerning the Alabama portion of the Soto route, the Commission collaborated with members of the Alabama Anthropological Society, formed in 1909. A specific location for Mabila was not proposed but the Commission did concur that the site was likely located somewhere between the Alabama River and Tombigbee River in Clarke County or in extreme southern Marengo County. That hypothesis has not been validated.

In 1977, another researcher, George Lankford, rekindled interest in the Alabama Soto route by proposing hypotheses of the locations of several Native towns in the Warrior River drainage possibly encountered by the Soto Expedition.

While the subsequent archeological testing of his hypotheses proved to be negative, more importantly, Lankford's inquires sparked decades of research concerning 16th-Century Spanish expeditions in the Southeast.

Based largely on Lankford's stimulus, archeological searches for Mabila and Nanipacana began in earnest during the 1980s and 1990s. The Alabama Historical Commission, University of Alabama, Alabama-Tombigbee Regional Commission and Mobile Historic Development Commission sponsored 16th-Century Spanish archeological research in the Warrior River and Alabama River drainages.

During the 1980s, a State Commission was formed in Alabama to study the Soto Expedition. The 1980s group was known as the Alabama De Soto Commission and was based at the University of Alabama. The commission was modeled after the "Federal De Soto Expedition Commission" of the 1930s. Meetings were held and a series of "Working Papers" was published by the commission. The subsequent Soto Expedition route debates could be described as … "lively."

A Federal De Soto Commission was also established during the 1980s including involvement by the U.S. Congress. The decision of the State and Federal Commissions along with the National Park Service was that the Soto Route would be marked through portions of the Southeast with roadside signs. The route was also widely publicized by the printed word through federal and state outlets, public news sources, and textbooks.

The decision of the State and Federal De Soto Commissions along with the National Park Service to "officially" endorse one specific Soto route prior to archeological testing is a perfect example of Dr. Swanton's prophetic statement published in 1939:

Nothing is made correct because it is called "Official."

The "Official" Soto route has now been archeologically surveyed and tested for over twenty years. Only one 16th-Century Spanish artifact has been found in the proposed area of Mabila and Nanipacana. It appears that Dr. Swanton was correct in his assertion.

Numerous research articles and books resulted from the Contact Period research during the decades of the 1980s and 1990s. Publication outlets included the Alabama-Tombigbee Regional Commission, Mobile Historic Development Commission, Alabama De Soto Commission, Smithsonian Institution, Florida Anthropologist, Pensacola Archeology Lab, Soto States Anthropologist, the University of Alabama, and others.

The 21st Century brought renewed interest in the first European expeditions into what is now the current

United States. Researchers were still intent on discovering the 16th-Century Spanish sites of Mabila, Nanipacana, the Pensacola Bay Spanish Colony, and the Luna campsite on Mobile Bay.

One hypothesis was proposed that the Soto army went directly south after leaving the Talisi Chiefdom on the lower Tallapoosa River in the general Montgomery, Alabama, area. The hypothesis proposed that the Bay of Ochuse was not Pensacola Bay but Choctawhatchee Bay. Though sound in reasoning, archeological testing proved the hypothesis invalid due to a lack of known Spanish artifacts through the drainages of the Choctawhatchee River, Pea River, and Conecuh River in southern and southeast Alabama.

The University of Alabama (UA) also continued hypothesis testing of prospective Soto sites during 2006 and 2007 along the Alabama River drainage in the central Alabama counties of Dallas and Wilcox. A research group, from several organizations, was formed to devise hypotheses concerning the potential archeological locations of the Native towns and Spanish contact sites of Mabila, Nanipacana, and Piachi. Thus far, archeological field tests by the UA group do not support their proposed locations for the locations of Mabila, Piachi, and Nanipacana.

A new research group was formed in 2013 known as Contact Archeology Inc. The group is focused on archeological and historical research concerning the first contacts of Europeans and Native peoples in the Southeastern United States. An online journal website (archeologyink.com) was created to disseminate hypotheses and discoveries.

Archeological hypothesis testing is, currently, being conducted by Contact Archeology Inc. on the lower Alabama River, generally, between Monroeville and the junction of the Alabama and Tombigbee Rivers to locate the sites of Mabila and Nanipacana.

The following pages provide a synopsis of discoveries made over the years by numerous archeologists in that region. The discoveries are impressive and form a Native settlement pattern and distribution of 16th-Century Spanish artifacts that might, indeed, lead to the discovery of Mabila and the "Settlement of the Sacred Cross" at Nanipacana.

Figure 13: Field Survey 2016.
Contact Archeology Inc. Files.

Durant Bend Site, 1Ds1

(Cottier 1970; Curren 1984, 1992; personal communication, Charles Clark and Ned Jenkins, 4/19/16; Brannon 1948; Moore 1899; Sheldon 1974; Nance 1976; Curren 1984; Knight 2009)

Figure 14: Spanish Candleholder, Two Views. Compliments Charlie Clark. Contact Archeology Inc. Files.

This site was hypothesized as Piachi, the Native town of the Soto and Luna expeditions. The hypothesis stated that after crossing the Alabama river here they marched to Mabila in the Cahaba River drainage. Archeological survey and testing proved the hypothesis to be untenable due to the lack of Mississippian Period sites in the Cahaba River basin, the lack of a candidate for Nanipacana in the vicinity, and the mismatch of the Piachi description and 1Ds1. Nonetheless, the research at 1Ds1 is of interest.

Two local men found a brass candleholder dating to the 1500s at site 1Ds1 in Dallas County, Alabama (see photos above). The candleholder was found about 150 yards from the Alabama River. The brass candleholder is generally shaped like a capstan used on early sailing ships, hence the name "capstan-style candleholder." The actual candleholder is attached atop the base.

The bottom of the base is approximately 6 inches in diameter. The entire artifact is approximately 5 - 1/2 inches in height.

The candleholder was found in an aboriginal burial pit, which was oval in shape and measuring approximately 6 feet in length and 3 feet in width. The candleholder was lying on its side approximately 4 inches from the top of the skull. The skeletal remains were found about 3 feet below the present ground surface. The burial was extended on the back with the cranium pointing to the north. No data is available on the skeletal remains, which were in a good state of preservation. Other than the candleholder, the pit fill was reported as containing approximately 3 shell tempered, incised pottery sherds and several small chert flakes. Beneath the burial pit was culturally sterile yellow sand.

Two other aboriginal burial pits were found within 6 to 10 feet of the one containing the candlestick. One of these was an urn burial and one was a bundle burial with several small shell beads. All three burials were associated with daub that may have been part of an aboriginal structure. Based on pottery types, two Mississippian Period archeological phases are known from the site, Furman Phase and Alabama River Phase, which date from approximately 1450-1650 AD.

The site was first reported by Clarence B. Moore in 1899. Moore was conducting an archeological survey and testing project of the Alabama River by steamboat. The research project was conducted under the auspices of the Academy of Natural Sciences of Philadelphia.

Moore discovered that a large flood in 1886 exposed "… human bones, earthenware vessels, whole and in fragments, and various other objects of aboriginal make." Local people later made numerous visits to the site and removed many artifacts, mostly by using probe rods to find pottery vessels with burial remains. We know nothing about the types, numbers, or time periods of those artifacts.

Moore and his steamboat team obtained permission to excavate on portions of the site. They recovered some 48 burials, some in burial urns, some bundled and extended. Hundreds, if not thousands, of artifacts were recovered.

Archeologists returned to the site in 1970 for excavations. The project was sponsored by the Archaeological Research Association Inc. and the University of Alabama Birmingham. The Native mound at the site was determined to be from the Woodland Period. Protohistoric deposits were also found on a portion of the site. Postholes, refuse pits, and burials (mostly infants) were found during the excavations. The area of Moore's excavations was also discovered.

In 1982, another excavation was conducted at the site. A cooperative effort between the University of Alabama Museum of Natural History and Auburn University Montgomery was conducted during the summer of that year. A Protohistoric Period structure and several burials were discovered.

The only known 16th-Century Spanish artifact found by any of these considerable excavations was the one candleholder recovered by the local men. 1Ds1 was a large Native habitation site during the Woodland and Mississippian periods. It was likely occupied during the Soto and Luna expeditions.

The Spanish candlestick is extremely rare. It is remarkably similar to another found on the lower Alabama River in Baldwin County. They are the only two found in the eastern United States. It was buried with a Native, likely a high-ranking person of the town. Was it obtained by the Native peoples through trade, theft, booty? We do not know. Due to the lack of Spanish artifacts, we do know that 1Ds1 was a Native town but was likely not a 16th-Century Spanish settlement, battle site, or the Piachi river crossing.

White Oak Creek Site, 1Ds53

(Chase and Herman 1969; Chase 1982; Jenkins and Paglione 2008; Knight et al. 2009, 2015)

In 1969, an archeologist with the Montgomery Museum of Fine Arts was conducting a survey searching for sites associated with the Soto Expedition through Alabama in 1540. A collector guided him to the White Oak Creek Site in Dallas County on the east side of the Alabama River.

Surface collections and limited excavations were conducted during 1969. Additional excavations were accomplished in 1974 sponsored by Hammermill Paper Corporation.

Based on the archeological excavations, it was discovered that the site had been occupied continuously from the prehistoric early Archaic Period into the 19th Century. Located along a historic trail that was likely a Native trail for hundreds or even thousands of years, the site was important to archeology in the Southeast. No evidence of a 16th-Century Spanish presence was found.

The White Oak Creek Site, once again, became the subject of Soto Expedition research in 2006. A group of researchers was formed in 2006-2007 to search for the site of Mabila. The group consisted of members from the University of Alabama, the University of South Alabama, Auburn University, Alabama Historical Commission, USDA-Natural Resources Conservation Service, and Troy State University. The group returned to the White Oak Creek Site and conducted excavations to test their hypothesis that the Native town of Piachi was located there.

The excavations consisted of eleven 2x2 meter units. No detailed analysis of the artifacts recovered has been published to date, but the research group concluded that the site was not Piachi due to the small Native occupation during the time of the Soto Expedition. Piachi was reported as a larger, important Native town by the Soto and Luna chroniclers.

The research group also tested their hypothesis that the battle site of Mabila was located in a region west of the Alabama River in Dallas or Wilcox counties. The field tests also proved to be negative.

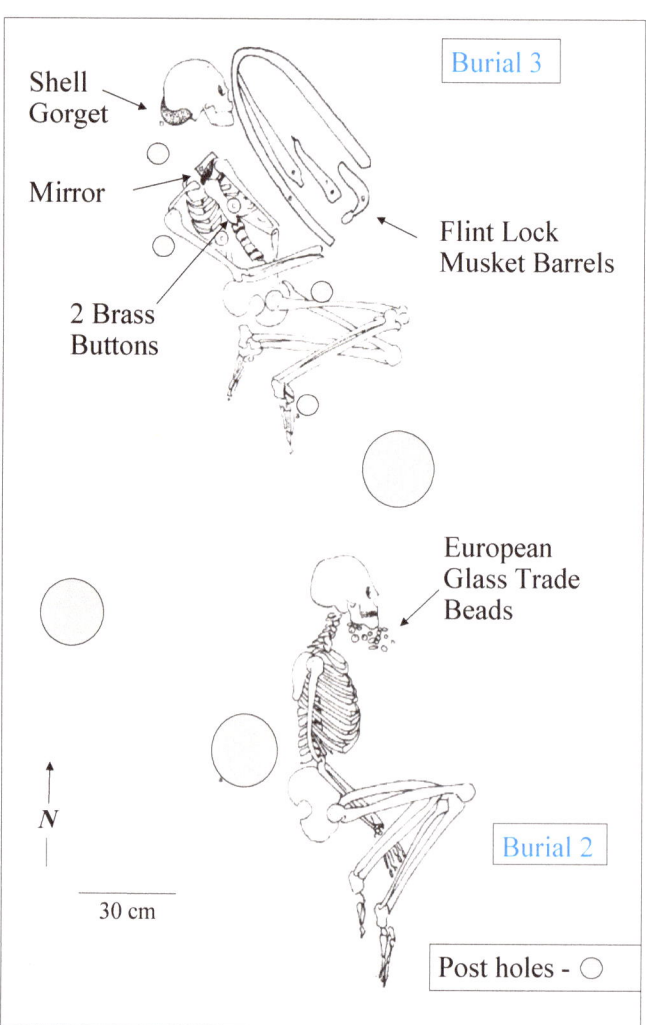

Figure 15: Two Native Burials with European Artifacts Likely from the 1700s. Excavated in the 1970s. Drawing by Ann Alford from David Chase Field Notes. Contact Archeology Inc. Files.

Crenshaw Site, 1Bu35

(Jenkins and Paglione 2008; personal communication, Charlie Clark and Ned Jenkins)

1Bu35 was recorded by chance. A United States Department of Agriculture official happened to be giving a presentation on archeology at a Butler County gathering of local people in 2004. A man from the audience came up to him after his lecture and invited him to examine artifacts from his property. It turned out that 16th-Century Spanish artifacts were among the collection as were Native Mississippian and Woodland Period artifacts. The site is located on private property beside a tributary stream 20 miles east of the Alabama River.

16th-Century Spanish glass trade beads were among the collection. The bead types included 1 Nueva Cadiz, 1 faceted chevron, and at least 6 blue spherical and oblong-shaped beads. Native shell beads far outnumbered the Spanish glass beads. Over 200 shell beads were found on the site along with a shell ear ornament. The high percentage of Native shell beads compared with the very low percentage of Spanish glass trade beads can be an indicator of early European contact. A brass disk made from European materials was also found at the site. A small hole was drilled in the center, perhaps by Natives (see photos next page). Based on field investigations by the Alabama Historical Commission, it appears that the site is the remains of a plowed-down Native burial mound.

The topography of the area of the site is unique and dramatic with high hills dropping radically to narrow stream valleys, reminiscent of the description of the Native town of Piachi which the Soto chroniclers described as located in rugged terrain.

However, this Butler County site does not appear to be the site of Piachi. The Soto writers clearly noted that Piachi was on a large river. Conversely, The Crenshaw Site is approximately 30 miles east of the Alabama River.

The site is, however, an important clue to our current study in that it was likely located along a major Native trail that ran from the Montgomery area to the lower Alabama River. The trail became known in the 1800s as the Federal Road. One of a number of hypotheses could be that the Spanish trade items found at 1Bu35 could have come from the Luna expeditionary force sent inland to Native Coosa in the north from the Spanish settlement at Nanipacana in the south. The European artifacts found at the site almost certainly came from the Luna and/or Soto expeditions and, as such, makes the site a key in Contact Period research.

Figure 16: A Portion of the Crenshaw Site. Photo Courtesy Ned Jenkins.

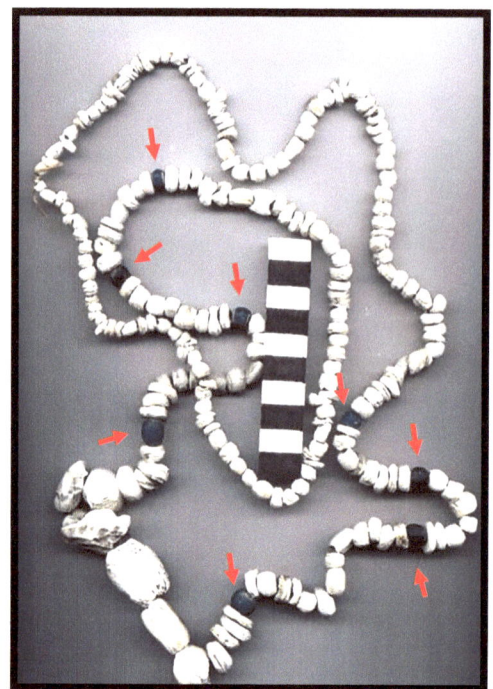

Figure 17: 1Bu35, Site View and Artifact Sample: Native Shell Beads, European Glass Beads (red arrows) and Brass Disk. Courtesy, Ned Jenkins, and Charlie Clark.

Liddell Site, 1Wx1

(Sears 1959; Bozeman 1963; Cottier 1968, 1970; Sheldon 1974; Hill 1979, Curren 1984)

The Liddell Site was one of the largest Native towns on the Alabama River during the Protohistoric Period. During the late 1500s and throughout much of the 1600s many Native people lived on this site. They were living in hard times. The diseases inadvertently brought to this land by the Spanish Expeditions of Soto and Luna hit the indigenous people hard and brought about a drastic change in their cultural identity. Their socioeconomic, political, and personal lives were dramatically changing. For the purposes of this current Contact Period study, it is not necessary to go into the details of this cultural change.

The artifacts and features at the Liddell Site reflect these changes. The excavation of numerous subsurface features at the site provide us with conclusive evidence of the relevancy of the site to the Contact Period.

A sample of burials included: 21 urn burials, 11 inverted vessel burials, 1 bundle burial, and 1 flexed burial. Only 1 burial feature contained European trade beads. The beads included 4 white and 7 turquoise blue dating to the 1600s.

Suffice it to say, that this site is neither the remains of Piachi or Nanipacana of Soto and Luna Expeditions. None of the criteria for those sites are present at the Liddell Site.

Native Bundle Burial   Native Structure

Figure 18: Liddell Site Excavations, 1968.
Courtesy University of Alabama
Office of Archaeological Research.

Goat Pasture Site, 1Wx12

(Cottier 1968; Sheldon 1974; Curren 1984)

This Wilcox County site has not been suggested as any specific Soto/Luna contact site but possible late 16th-Century Spanish artifacts have been found on the site, thus, a summary of the site is appropriate for this exposé.

Excavations at 1Wx12 were conducted in 1965 by the University of Alabama. It is a multicomponent site. Based on pottery types, the largest occupation occurring during the Protohistoric Period. Thirty-one features were excavated, all dated to the Protohistoric (late 1500s to mid-1600s). Burials, refuse pits, postholes, and a possible structure were discovered.

Four burials were excavated at the site, all sub-adults. Three burials were found in ceramic urns and one was extended. The extended child burial (7-11 years of age) contained the following objects found on the upper chest: 2 shell pins, 1 small circular shell pendant, and 19 European glass trade beads. One of the glass beads was small and white and 18 were small turquoise-blue beads.

The bead types have a long time range but were likely from the second half of the 1500s to the first half of 1600s. This might, possibly, indicate contact with the Luna Expedition, but it is not a certainty.

It is a certainty that 1Wx12 is not the site of Nanipacana or Piachi. None of the criteria for these sites, listed earlier, are present at this site.

Figure 19: Goat Pasture Site Excavations, 1968. Images Courtesy University of Alabama Office of Archaeological Research.

Furman Site, 1Wx169

(Moore 1899; Jenkins and Paglione 1980;
Curren 1984; Regnier 2005)

The indefatigable Clarence B. Moore first reported this site in 1899. Steaming upriver on his trusty steamboat, the "Gopher," the research party located the site on the Alabama River in Wilcox County. Mounds, Mississippian Period pottery vessels, and Native burials were reported.

Archeologists returned to the site in the 1980s during a survey funded by the Alabama Historical Commission and Auburn University. They recorded the site in the state site files as 1Wx169.

Archeologists from the Alabama-Tombigbee Regional Commission (ATRC) conducted excavations at the site later in the 1980s to test the hypothesis that the site could be the site of Nanipacana mentioned in the Luna Expedition writings. Fifty excavation units of varying sizes were dug at the site. The subsurface Native archeological features discovered at the site included: refuse pits, fire hearths, postholes, structures, and burials. Sixteenth-Century Spanish artifacts were not found at the site, indicating the likelihood that the Furman Site, while a very important Mississippian archeological site, did not figure into the first Spanish contacts in the region.

Figure 20: 1Wx169, Mapping Team and Aerial View of Site. Contact Archeology Inc. Files.

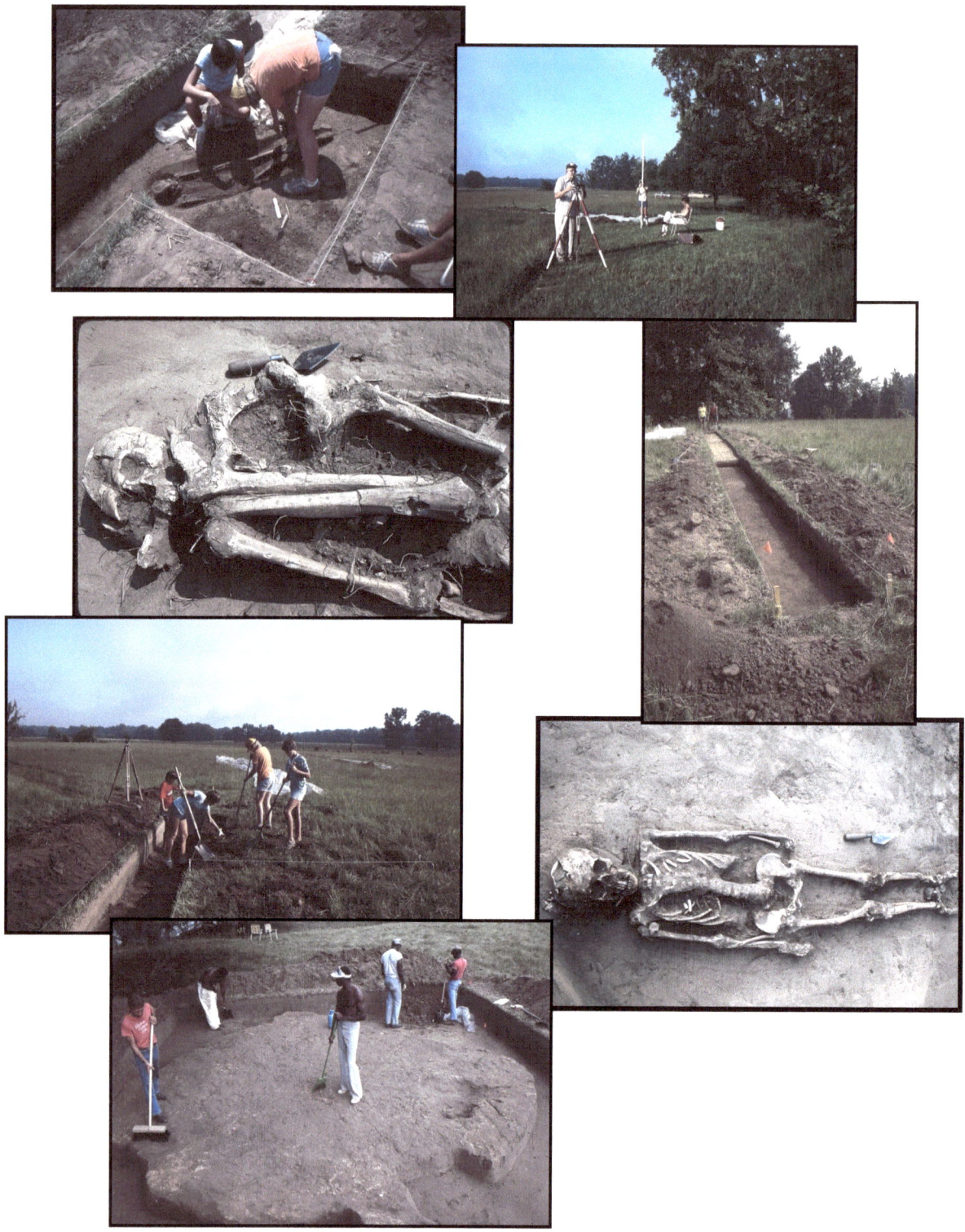

Figure 21: 1Wx169, Field Photos.
See Curren 1984 for details.
Contact Archeology Inc. Files.

26

Figure 22: 1Wx169, Field Photos.
See Curren 1984 for details.
Contact Archeology Inc. Files.

Claiborne Dam Site, 1Mn6

(Graham 1967; Curren 1984, 1987,1992; personal communications, Richard Williams 1991; Nicholas H. Holmes Jr. 2015)

It has been hypothesized that 1Mn6 is the site of the Soto army river-crossing site at the Native town of Piachi. Most or all of the site has been destroyed by the U.S. Army Corps of Engineers. There were, however, archeological investigations prior to the destruction.

The site was first found by a local man in the 1960s. It is located in Monroe County adjacent to the Alabama River on a narrow floodplain backed by rugged hills.

During 1965-66 the University of Alabama (UA) conducted excavations at the site prior to construction of a lock and dam by the U.S. Army Corps of Engineers. Several sites were recorded on the east side of the river and excavations were conducted. The sites are now considered one site, 1Mn6. The site is large, 1/4-mile long, 100-200 yards wide. Archaic, Woodland, and Mississippian components were present including subsurface features.

During the construction of the lock and dam numerous features and artifacts were recovered by local people. An archeologist from the Montgomery Museum of Fine Arts (MMFA) was present at times and made a collection with field notes.

Additional archeological research was conducted during the 1980s and 1990s on the lower half of the Alabama River under the auspices of the Alabama-Tombigbee Regional Commission, Mobile Historic Development Commission, and the Pensacola Archeology Lab.

During this current research project, 1Mn6 was revisited. Photographs were made. Due to the massive lock and dam construction, it is likely that little, if any, of the site remains. Yet, it is possible that deep cultural features may have survived.

We plan to locate and reexamine the UA and MMFA collections of Mississippian Period artifacts and potential 16th-Century Spanish artifacts. The site has been hypothesized as a potential location of the Native town of Piachi visited by both the Soto and Luna expeditions. Based on current data, the hypothesis is disproven. No Spanish artifacts have been recorded from the site.

Figure 23:
1Mn6, Claiborne Dam.
See Curren 1984 for details.
Contact Archeology Inc. Files.

Nancy Harris Landing Site, 1Mn183

(Moore 1899; Sheldon 1974; Curren and Majors 1984; Curren and Lloyd 1987; Curren 1987)

This Mississippian Period site is one of the largest on the Alabama River. This fact was recognized as early as 1899. The site is approximately 1/2 mile long and approximately 1/2 mile wide (eroded by floods). Consequentially, it has been hypothesized as the site of Nanipacana, noted in the Luna papers as the largest Native town in the region. It has not been archeologically proven as such but more field investigations are scheduled for the 2016 season.

When Professor Clarence B. Moore and his team from the Philadelphia Academy of Sciences arrived at the site by steamboat in 1899, they were told by locals that the large "freshet" (flood) of 1886 had exposed many artifacts over a large area of the site. The artifacts included pottery vessels "broken and whole," ceramic pipes, and human bones.

Moore and his team, thirteen years after the flood, still found ceramics, effigies, and human bone on the surface of the site. Moore was used to finding mounds containing dramatic in-situ "whole" artifacts. These scatters of broken artifacts on the surface did not impress him. Archeologists of the present day would, indeed, be impressed. After using probe rods in the area and finding "nothing of significance," Moore concluded that the site had been mostly washed away by the 1886 flood.

Archeological field teams from the Alabama-Tombigbee Regional Commission returned to the site in 1983, other teams in 1991 from the Pensacola Archeology Lab, and in 2015 from Contact Archeology Inc.

Thus far, the necessary archeological criteria to prove that the Nancy Harris Landing Site is Nanipacana have not been met.

Figure 24: 1Mn183 River View.
See Curren 1992 for details.
Contact Archeology Inc. Files.

It was discovered that the entire site was not washed away by the 1886 flood. A large portion had, indeed, been impacted as evidenced today by flood scarring on the eastern portion of the site. However, an intact Native midden was discovered at places along the eroding river bank. Midden was exposed at intervals for approximately 1/2 mile along the river.

The midden thickness varied from 12-18 inches (30-45 cm.) and was buried by culturally sterile sand river deposits, a thickness of approximately 18 inches (50 cm). Woodland and Mississippian artifacts, primarily ceramics, and at least one feature (cremation burial) may have come from the site. The largest occupation was during the Mississippian Period.

Contact Archeology field teams made numerous trips to the site in 2015. Shovel tests and surface collections confirmed the Native occupation periods at the site but no 16th-Century Spanish artifacts were recovered. The artifacts have been recovered by local people and the archeological field teams. A total of 325 Mississippian Period pottery sherds have been found by Contact Archeology members at the site. The diagnostic artifacts indicating a Mississippian Period occupation at the site are: Pensacola Incised vars. *Gasque, Jessamine, Bear Point*; Moundville Incised vars. *Moundville, Snows Bend*; Mound Place Incised *var. Walton Camp*; Bell Plain, *var. Hale*; D'Olive Incised, *var. Arnica.*

A rather unique geologic deposit appears at the up-river edge of the site. The regional deposit is known worldwide to researchers. The deposit dates from the Eocene and contains abundant marine fossils. This small portion of the regional formation, present as a ridge spur, forms a small hill adjacent to the main archeological site and is clearly seen from the river. It may be a clue to the reference in the Spanish writings of the Luna Expedition. Reportedly, a hill was present at the Native town of Nanipacana. The Native town name translates as "Hilltop." Field teams from Contact Archeology Inc. made several trips to the site during 2015 and intend to return to the site in 2016. Private collections are also being examined. Thus far, the archeological criteria necessary for the site of Nanipacana at the Nancy Harris Landing Site (1Mn183) have not been met.

Figure 25: Hilltop at 1Mn183 formed by Geologic Eocene Deposits. Contact Archeology Inc. Files.

Figure 26: 1Mn183. River Views, Pottery Vessel, Shovel Test.
Details in Curren 1994.
Contact Archeology Inc. Files.

Broughton Temporary Site #1

*(Field Notes, Caleb Curren 1983-2015; personal communication,
Nicholas H. Holmes Jr. and Ann Bedsole Holmes 2015)*

This site is located in Clarke County on the opposite side of the Alabama River from 1Mn183. As far as Contact Archeology Inc. is concerned, it is a Mabila candidate. We plan to test the hypothesis with fieldwork during the 2016 campaign.

The site was first discovered during our Soto/Luna site investigations during the 1980s. It contains Woodland and Mississippian components and possibly Archaic as well. A site form will be submitted to the Alabama State Site Files when more fieldwork determines the actual number of sites and specific locations with which we are dealing.

The area contains several small sites or one or two larger sites. None are nearly the size of nearby 1Mn183. There are five other sites dispersed across the flood plain beyond a small creek flowing to the Alabama River. They all seem to have similar components, although, this is not a certainty due to the paucity of shovel testing accomplished at any of the sites.

The sites are buried under 8-20 inches (20-50 cm.) of sandy river sediment. There is a small, natural pond in the area, likely formed as a sinkhole due to the limestone substrate in the region. Other sinkhole ponds are also located in the area.

When first investigated, the area was mostly plowed fields. Currently, it is mostly planted pine trees.

A possible Native mound has also been reported in the area of these sites. The site is considered a potential Mabila candidate. To date, it has not confirmed as such.

We plan to return to the site during the 2016-17 season to gather further archeological data concerning possible associations with the Contact Period.

*Figure 27: Satellite Image of the
Broughton Site Area.*

Ideal Cement Site, 1Mn57

(Brock, Fuller, and Lau 1978; Curren 1987, 1992)

The site was first reported by archeologists from the University of South Alabama (USA) in 1978 during surveys and excavations prior to a 4,000-acre limestone quarry expansion project. Excavations at the site consisted of ten 2m x 2m units and nine 1m x 1m units. The soil was screened through 1/2-inch mesh wire screens. A bulldozer cleared the site prior to the excavations. The site was multicomponent, and included considerable Mississippian Period artifacts. A total of 626 small pottery sherds were recovered, of which 537 (87%) were shell tempered (487 plain, 47 incised, 3 punctate).

The site was large, approximately 250-300 yards long and 100-150 yards wide. Its location is unusual for a Mississippian Period site in this area. It was situated on a hilltop plateau approximately 100 feet above the east side of the Alabama River. More often, sites of the period are located in the floodplains of the rivers in this region due to the Native's reliance on fertile, river bottom soils for agriculture.

Could the site have been the location of the 16th-Century Native town of Piachi or Nanipacana? The Soto writers reported Piachi as situated high above a large river. The Luna writers reported the name of their inland settlement was established at the Native town of Nanipacana, translated as "Hilltop." The lack of Spanish artifacts at the site indicates that this hypothesis is unlikely. Unfortunately, a more detailed search for features at the site was not accomplished.

It was recommended by USA archeologists that monitoring of the planned quarry mining expansion at the site be implemented. The archeologists even offered to assemble a volunteer field team for that purpose. The Ideal Cement Company rejected the generous USA proposal. The site was later destroyed by the quarry expansion (personal communication, Richard Fuller 1992).

Figure 28: 1Mn57, Location Map.

Jewitt Landing Site, 1Ck55

(Personal communication, Louis Finlay and Ben and Calvin Syphrit 1983; Fuller, Silvia, and Stowe 1984; Curren and Majors 1987; Curren 1992)

The year was 1918. The lower Alabama River region was even wilder than it is today. A young man was working with a timbering crew rafting logs down the river. He noticed what seemed to him to be human bones near the top of a high bank of the river. He managed to get off the raft of logs and scramble up the steep riverbank to the feature. He found "a skeleton in a pot." A pottery vessel was inverted over human long bones and a skull with a Spanish silver coin lying amongst the bones. He removed the entire burial, put it in a box, and took it home. Over the years, the bones and pot were lost but the coin was kept.

The discovery is a major clue to the studies of the Contact Period in Southwest Alabama. It is a 16th-Century silver Spanish coin, a little larger than a modern U.S. quarter (approx. 1 inch). It was hand cut from a sheet of silver. A hole was drilled on one edge with wear marks from the necklace possibly rubbing on the chest of the owner, a Native who wore it in life and took it to the grave.

The marks on the coin unequivocally identify it as 16th-Century Spanish in origin. The coin was minted in Mexico City between 1554 and 1570. It was likely brought to the lower Alabama River by the Luna Expedition of 1559. It may have been traded to the Natives or taken off a dead Spaniard.

Although this Clarke County site was likely occupied during the time of the Soto and Luna Expeditions, it is not Mabila or Nanipacana. It is a breadcrumb along the trail of the Luna Expedition odyssey.

The site was revisited by Contact Archeology field teams during the 2015 research season. Contacts with local landowners were made. We intend to follow up with those contacts during the 2016 campaign.

Figure 29: The Syphrit Coin.
Obverse and Reverse Sides.
Contact Archeology Inc. Files.

Doctor Lake Site, 1Ck219

(Fuller, Silvia, and Stowe 1982; Curren 1987; Curren and McKenzie 1988; Curren 1992; Thompson 1998, 1999; Mikell and Little 1999; Little and Harrelson 2005)

This site is located on an oxbow lake approximately midway between the Alabama and Tombigbee in southern Clarke County. The site was first recorded in 1982. Additional archeological excavations were conducted in 1984, 1986, 1987, 1988, and 1991. The sponsoring agencies included the Alabama-Tombigbee Regional Commission, The Mobile Historic Development Commission, and the Pensacola Archeology Lab. The site was hypothesized to be the battle site of Mabila, hence, the repeated testing at the site.

The excavations were numerous and included 45 excavation units sized from 1x1 m. to 2x10 m. to 2x40 m. The units were hand dug and screened through 1/4-inch wire mesh. Units 22-34 were also hand dug after overburden was mechanically removed. These units were placed adjacent to a dirt road across the site. The plowzone in the river-born overburden ranged in thickness from 10-20 cm. (4-8 inches) and the occupational midden ranged from 5-90 cm. (2-36 inches).

The excavations resulted in nineteen features (burials, refuse pits, fire hearths) and 56 postholes, including 23 large postholes from a Mississippian Period burned palisade wall section adjacent to the aforementioned dirt road. The Native ceramic sherds from the site totaled 4,266, mostly shell tempered, **dating from the Mississippian Period and into the 1700s.**

Mississippian Period pottery types included: Pensacola Incised *vars. Gasque, Jessamine, Bear Point*;

Moundville Incised *vars. Moundville, Snows Bend, Douglas*; D'Olive Incised vars. *D'Olive, Arnica, Mary Ann, Dominic*; Salt Creek Cane Impressed *var. Salt Creek; Carthage Incised*; Bell Plain; Mississippi Plain.

The closed-context European artifacts recovered came from the twelve burial features. They included six bundle burials, one extended burial, and five undetermined burials due to bone deterioration. The European artifacts came from four of the burials: an iron "C"-shaped bracelet with Feature 1; one spherical purple and eight spherical turquoise blue glass trade beads with Feature 9; and two unidentified iron objects from Feature 10 (see Curren 1992 for details).

In summary, site 1Ck219 was occupied before, during, and after the Soto and Luna expeditions (Woodland, Mississippian, and Historic Periods). The site exhibits certain traits of the Native town and battle site of Mabila described in the Soto Expedition. It is a small, narrow town site adjacent to a "pond" (oxbow lake in this case). It had a palisade wall that was burned.

However, diagnostic 16th-Century Spanish artifacts and features dating to the Soto Expedition have not been found at the site, despite rather extensive excavations. The in-situ European artifacts were found with four of the Native burials. None of the artifacts dated to the Soto Expedition.

In addition, palisaded Native towns, some burned, have been excavated throughout the Eastern United States. One burned Native palisade wall does not prove that it's the site of Mabila.

When all these data are considered, it must be acknowledged that 1Ck219, while a good candidate for Mabila, has not exhibited the necessary criteria for the actual battle site itself.

Additional archeological investigations were conducted at 1Ck219 during the late 1990s and early 2000s by Jacksonville State University, the Alabama Historical Commission, and the Mobile Historic Development Commission. Magnetometer surveys and test excavations were conducted. That research effort, like the previous ones, did not discover necessary evidence for the location of Mabila at 1Ck219. Despite extensive excavations, 16th-Century Spanish artifacts were not discovered at the site. Very likely, the site is not the location of the Native village and battle site encountered by the Soto Expedition in October of 1540.

Figure 30: Doctor Lake.
Field Photos.
Details in Curren 1992.
Contact Archeology
Inc. Files.

Choctaw Lake Site, 1Ck218

(Picket 1851; Curren 1992)

Archeologists from the Alabama-Tombigbee Regional Commission (ATRC) conducted archeological excavations at this site during the summers of 1983-85. The Alabama historian Albert Picket proposed in 1851 that the battle site of Mabila was on a bluff near this site. ATRC investigated the bluff site to no avail. No significant Mississippian town sites or Spanish artifacts were present, only a Woodland Period campsite and a Civil War gun emplacement.

We then investigated an interesting area adjacent to the bluff which had the potential of being a Native archeological site. Through shovel testing in 1983 that proved to be the case.

The site is primarily a Mississippian Period village site, likely occupied during the Soto and Luna expeditions. The site size is approximately 9 acres. It is adjacent to the Alabama River and an oxbow lake abuts the site on three sides.

The occupational midden at the site is covered by a 6-12 inch (15-30 cm.) thick layer of alluvial, culturally sterile, dense clay overburden. The Mississippian midden thickness varied from 4-12 inches (10-30 cm.).

The results from 32 excavation units at the site included: a wall trench on the southern border of the site indicating a likely palisade, 14 features, 29 postholes, 800 shell tempered Mississippian Period sherds (total sherd count 865), and 4 shell tempered bottles found in the only burial encountered. Mississippian Period ceramic types included: Mississippi Plain, Pensacola Incised *vars. Gasque, Bear Point*; D'Olive Incised; Moundville Incised *vars. Moundville, Snows Bend*; Mound Place Incised *var. Waltons Camp*; Bell Plain *var. Hale*; Carthage Incised. Probable structures were present on the site as evidenced by widespread and consistent small daub fragments and at least one line of postholes.

The site was considered a possible Mabila candidate at the time. The lack of 16th-Century Spanish artifacts does not support that hypothesis. Nonetheless, the discovery of the site and subsequent excavations added considerable contributions to the knowledge of the Mississippian Period of the lower Alabama River. Images of the excavations at the site follow. Archeological details of the site are found in Curren 1992.

Figure 31: Choctaw Lake Site River View. Contact Archeology Inc. Files.

Figure 32: 1Ck218, Field Photos.
Details in Curren 1992. Contact Archeology Inc. Files.

Figure 33: Choctaw Lake Site Field Photos.
Details in Curren 1992.
Contact Archeology Inc. Files.

Driesbach Lake Site, 1Ck263

(Curren 1992; Little and Curren 1990)

The site location was predicted and found in 1988 using known Mississippian Period settlement patterns and topographic maps. The project was administrated through the Mobile Historic Development Commission and the Alabama-Tombigbee Regional Commission.

A total of 101 excavation units were completed during the field seasons of 1988-89. The size of the units included 40x40 cm. shovel tests, 2x2 m. units, and one unit measuring 3x4 m. The minimum depth of the units was 45 cm. The excavation units were dug at intervals along a relic river-channel lake adjacent to the site for several hundred meters.

A series of small sites, probably house sites, were found with the excavations. Some dated to the Woodland Period and some to the Mississippian Period. They may have been isolated hamlets or even a small town.

A total of 1,363 pottery sherds were recovered from the excavations, 1,073 being shell tempered along with numerous, large fragments of cane-impressed daub. Mississippian Period ceramic types included: Pensacola Incised *vars. Gasque, Jessamine, Bear Point*; Moundville Incised *var. Snows Bend*; Mound Place Incised *var. Waltons Camp*; D'Olive Incised *var. Mary Ann*; Salt Creek Cane Impressed *var. Salt Creek*; Bell Plain; Mississippi Plain *var. Warrior*.

Features and postholes were present. Small amounts of lithics and animal bone fragments along with four carbonized corncobs were also recovered. Spanish artifacts from the 16th-Century were not present in the collection. Due to the considerable amount of excavations and the lack of 16th-Century Spanish artifacts, the site is likely not Mabila. A sample of field photos follow.

Figure 34: Driesbach Lake Site.
Lowland Forest Terrain.
Contact Archeology Inc. Files.

Figure 35: Driesbach Lake Site, Field Photos.
Details in Curren 1992.
Contact Archeology Inc. Files.

Singleton Site, 1Ck41
(Curren and McKenzie 1988; Curren 1992)

This 7-acre village site is located on a small relic river-channel lake between the Alabama and Tombigbee Rivers. The site was first discovered by an archeological team from the Alabama-Tombigbee Regional Commission in 1984 when local collector, Ben Griffin, guided the team to the site. The site was tested periodically from 1984 to 1987.

A total of eleven 2x2 m. excavations were completed at 1Ck41. A dense sedimentary clay layer, 10-15 cm. thick, covered a 10-30 cm. thick midden at the site.

A total of 5,277 sherds tempered with sand, grit, and grog were recovered along with 3,745 shell tempered sherds. Postholes and features were encountered. Diagnostic Mississippian Period ceramic types included: Pensacola Incised *var. Gasque and Jassamine;* Moundville Incised *vars. Moundville, Snows Bend;* Mound Place Incised *var. Waltons Camp;* D'Olive Incised *vars. Arnica, Dominic, Mary Ann;* Carthage Incised; Salt Creek Cane Impressed *var. Salt Creek;* Bell Plain *var. Hale;* Mississippi Plain *var. Warrior.*

No 16th-Century Spanish artifacts were recovered from the site. The criteria for the site of Mabila have not been met at the Singleton Site.

Figure 36: Singleton Lake Site. Terrain and Vegetation. Contact Archeology Inc. Files.

Little River Site, 1Mn227

(Moore 1899; Jenkins and Paglione 1980; Fuller, Silvia, Stowe 1984; Little, and Holstein 1989; Little and Curren 1991; Curren 1992)

This site was a Native village and burial mound on the east side of the lower Alabama River in Monroe County. Artifacts indicate that it was likely occupied before, during, and after the Soto and Luna expeditions.

Excavations at the site started with the Philadelphia Academy of Sciences expedition headed by Clarence B. Moore. Following Moore's investigations came a 1980 survey by the Alabama Historical Commission. The survey team could not reach the site due to floodwaters of the river.

Next came research by the Alabama-Tombigbee Regional Commission (ATRC) in the 1980s. Next came a 1980s survey by the University of South Alabama, then, again in 1988 by the ATRC. The most recent archeological excavations at the site came in 1991, sponsored by the Mobile Historic Development Commission.

Based on all of these investigations, it was determined that the village size is approximately 750 feet long x 400 feet wide. A plowed-down burial mound was recorded by Moore as approximately 50 feet in diameter and 2 feet high. The original height of the mound before plowing is unknown.

Five burials were found in the mound by Moore. One in particular is most pertinent to the Contact Period. The feature contained multiple burials and included:

- "**Many Glass Beads**"
- **Pearl**: 1, perforated
- **Shell Gorgets**: 2, circular
- **Shell Beads**: 6, large
- **Bead**: 1, sheet copper
- **Pins**: 3, small shell
- **Pins**: 3, large shell
- **Ear Ornaments**: 9, shell disks
- **Disk**: 1, Native pottery
- **Burial Urn**: 1, incised, +flat cover
- **Rim Effigy**: 1, bird head

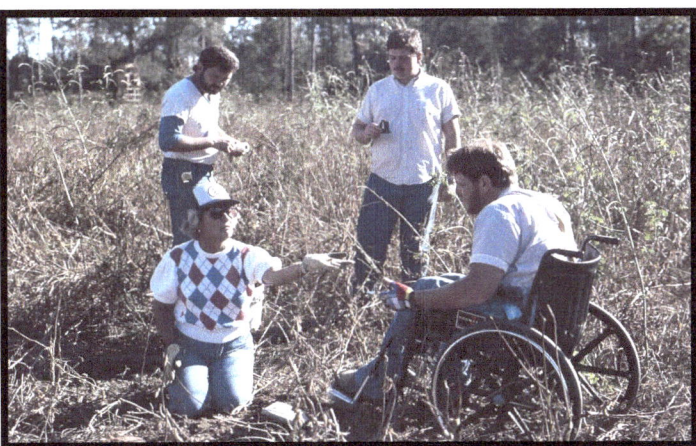

Figure 37: Little River Field Photo.
Details in Curren 1992.
Contact Archeology Inc. Files.

The ATRC researchers were able to track down a number of the Moore artifacts with the cooperation of the staff of the Smithsonian Institution.

The glass beads reported by Moore were not found in the Smithsonian collections. The "many glass beads" from the Moore expedition likely represent trade with Europeans but which Europeans … Soto, Luna, later French? Contact Archeology Inc. is currently searching for the beads in various museums.

The "perforated pearl" is also of particular interest. The chroniclers of the Soto expedition reported that, at the battle of Mabila, the Spanish lost their numerous freshwater pearls taken from the Natives earlier in their trek through the Southeast. Freshwater pearls are rare in archeological context. Moore noted that this pearl was the only one found by him in all of Alabama. Does this one pearl provide a hint that the battle site of Mabila is in this region of Southwest Alabama?

1Mn227 is on the wrong side of the river to be Mabila but the one pearl and, perhaps, the glass beads might provide clues to the proximity of the Battle of Mabila. The two photos below show excavations in the area of the plowed-down mound at the site. Archeological details are found in Curren 1992.

Figure 38: Little River Site Field Photo.
Details in Curren 1992.
Contact Archeology Inc. Files.

Pine Log Creek Site, 1Ba462

*(Stowe et al. 1982; Fuller et al. 1984; Curren 1987;
Little and Curren 1990; Curren 1992; Little and Harrelson 2005)*

The site is located on a small tributary stream on the east side of the lower Alabama River near the border of Baldwin and Clarke counties. It is composed of approximately five, low Native burial mounds. There is not a village present. The site encompasses some five acres.

Private collectors found the site with metal detectors in the late 1970s and/or the early 1980s. The individuals quietly dug at the site and removed numerous artifacts from the private property without permission. The Alabama Historical Commission and the University of South Alabama cooperated to recover the artifacts in concert with the landowner.

Mississippian Period ceramic types include: Pensacola Incised *vars. Bear Point, Jessamine, Pensacola, Perdido Bay*; D'Olive Incised *vars. Arnica, D'Olive, Mary Ann*; Moundville Incised; Bell Plain *var. Hale*; Mississippi Plain.

- **Native Pottery Sherds**: 4,292
    (shell tempered = 4,265)

- **Native Pottery Vessels**: 46
    (complete or partial, all shell tempered)

- **Native Copper, Arrow-shaped Objects**: 6
    (all copper, likely belonging to a headdress)

- **Native Stone Artifacts**: 58

- **Native Shell Beads/pins**: 72
    (for necklaces and/or hair or ear ornaments)

Figure 39: Small Sample of Pottery Vessels and Stone Ceremonial Axe from 1Ba462. Courtesy University of South Alabama.

The site is one of the most significant archeological sites of the early European Contact Period in the entire country. The impressive list of recovered European artifacts are :

- **Candlestick**: 1, brass
- **Container**: 1, brass
- **Gun Barrel**: 1, iron
- **Ladle**: 1, iron
- **Sword**: 4, iron fragments
- **Lance/Pike Head**: 1, iron
- **Knife**: 1, iron
- **Horse Bridle and Cheek plate**: 1, iron
- **Horseshoe**: 1, iron
- **Kettle Fragment**: 1, iron, reworked
- **Axe**: 1, iron
- **Chisel**: 1, iron
- **Wedge**: 1, iron
- **Sickle**: 1, iron
- **Spikes**: 4, iron
- **Unidentified Objects**: 8, iron
- **Trade Beads**: 4, faceted glass, seven color layers
- **Trade Beads**: 5, glass, blue
- **European Ceramic Sherd**: 1, Columbia Plain, reworked into a Native ear spool

1Ba462 is not the site of Mabila nor is it the Settlement of the Sacred Cross. However, the 16th-Century Spanish artifacts found in the Native burials in the five mounds at the site likely came from the Soto and/or Luna expeditions. Is the site a clue that the archeological sites of the Soto battle site and the Luna settlement are nearby?

The research strategy of Contact Archeology Inc. for the 2016-17 campaign includes following leads obtained from local people concerning other potential Spanish contact sites on both sides of the lower Alabama River from the Monroeville area to the vicinity of 1Ba462. The research is a multiyear project.

Figure 40: European Glass Trade Beads, 1Ba462. Contact Archeology Inc. Files.

Figure 41: Pine Log Creek Sample of European Artifacts. Details in Curren 1992. Images by University of South Alabama and Contact Archeology Inc.

"Site-X"

(site files, Contact Archeology Inc.)

Allegedly, there is a 16th-Century Spanish contact archeological site on the west side of the Alabama River relatively near 1Ba462. The site was found by two collectors during the 1980s. The men were searching for Mabila and plotting their version of the Soto route on maps.

According to the collectors, the site floods during portions of the year and is covered by 6" to 12" of dense, culturally sterile, alluvial clay. It was found by the collectors using several techniques which included infrared aerial maps, metal detectors, and night vision goggles. Hand tools were used to recover the artifacts. No screens were used.

Some of the metal artifacts found began to corrode once removed from the ground so the men needed to learn preservation techniques. They visited Jefferson Davis Community College in Alabama for that reason. During their visit to the college with their artifacts, a general inventory of the artifacts was made. The men allowed no photographs to be taken. They promised to revisit the college later. They never did. The notes and names taken during the visit have since been lost. The information contained in this document came from telephone interviews by Contact Archeology Inc. The artifacts viewed at the college and thought to have come from "Site-X" include:

- **Thimble**: 1, metal source unknown, very ornate
- **Glass Beads**: approx. 6, 1 round blue, 5 oblong blue and multicolored
- **Matchlock Gun Parts**: several, metal, all from matchlock mechanism itself
- **Hammer**: 1, iron, small, cobbler? blacksmith? carpenter?
- **Dagger**: 1, steel, good condition, T-shaped hilt, found inside a Native pottery vessel
- **Native Pottery Sherds**: many
- **Native Stone Game Stones**: several
- **Coins**: Spanish, several silver, 1 gold, 1 hole-drilled
- **Buckle**: 1, iron, large, round, bar across center
- **Crossbow Projectile Point**: 1, dense iron, well preserved, 1 badly eroded
- **Sword**: 1, steel, basket hilt, rapier-like blade, double-edged, fluted, found buried vertically
- **Gun Shot**: several, iron, eroded
- **European Majolica Ceramic Sherds**: several

The Contact Archeology Expedition of 2015-16: Hypothesis Testing on the Lower Alabama River

Artifacts from the 16th-Century Spanish expeditions of Soto (1540) and Luna (1559) have been found in the Alabama River drainage of central and southwest Alabama. Historical documents relative to the two expeditions have been examined. The original historic documents are located in the archives of Europe, Mexico, the Caribbean, and the United States. These historical and archeological explorations are major steps in our better understandings of the dramatic events in Southwest Alabama and Northwest Florida relative to the contacts of peoples from two continents in the 1500s. The events are unique in world history.

The 2015 research campaign of Contact Archeology Inc. in Southwest Alabama was multifaceted. The goals involved: historical documents research, archeological background research, property access permission, revisits to previously recorded archeological sites, accessing private artifact collections, and following leads on recent discoveries by local people. The goals were met.

The 2015 archeological expedition was designed to lay the groundwork for a multiyear investigation. We reestablished our research presence in the area following the intensive archeological studies during the 1980s and 1990s.

The 2015 campaign also determined that 16th-Century Spanish artifacts are not consistently distributed in the Alabama River drainage. Some areas of the drainage contain more Spanish artifacts than others. This is likely a significant clue to the locations of Native towns recorded in the Soto and Luna documents.

The region of the junction of the rivers of the Alabama, Tallapoosa, and Coosa near Montgomery contain a concentration of 16th-Century Spanish artifacts. The region of the lower Alabama River is also a region containing concentrations of 16th-Century Spanish artifacts. In contrast, the region between Montgomery and the lower Alabama River contains dramatically fewer numbers of Spanish artifacts.

Logically, the areas where the Spanish expeditions of Soto and Luna spent the most time would contain the most Spanish artifacts. Conversely, the areas where the Spaniards spent the shortest time would contain fewer numbers of Spanish artifacts.

Figure 42: Lower Alabama River from Contact Archeology Field Boat.
Contact Archeology Inc. Files.

The current hypothesis proposed by Contact Archeology states that the relatively large concentration of Spanish artifacts in the upper Alabama River / lower Tallapoosa River region near Montgomery represents the location of the Talisi Chiefdom encountered by the Soto Expedition.

The hypothesis also states that the lack of significant numbers of Spanish artifacts between the Montgomery (Talisi) region and the lower Alabama River region reflects shorter stays by the Soto and the Luna expeditions as reported in the Spanish writings.

The hypothesis also states that the relatively large concentration of Spanish artifacts in the lower Alabama River region represents the location of the Mabila Chiefdom encountered by the Soto Expedition. The hypothesis states that Nanipacana (The Settlement of the Sacred Cross) and Mabila are located in that region.

Densities of Native settlements could also be a clue to the discovery of Mabila and Nanipacana. Dense Native settlements were described by the Spaniards only at Talisi to the north and Mabila and Nanipacana to the south in the Alabama River drainage. The densest Native occupations in the Alabama River drainage are in the Montgomery region (Talisi Chiefdom?) and the lower Alabama River - Mobile Delta region (Mabila Chiefdom?).

The combination of 16th-Century Spanish artifact concentrations and Native settlement concentrations could prove to be a key to the discovery of Mabila and Nanipacana. The concentrations of the Spanish artifacts and the Native settlements is the core of the current Contact Archeology hypothesis. The validity of the hypothesis is yet to be proven or disproven.

Figure 43: Lower Alabama River Tributary Stream. Contact Archeology Inc. Files.

(Sample photographs of the 2015-16 fieldwork follow.)

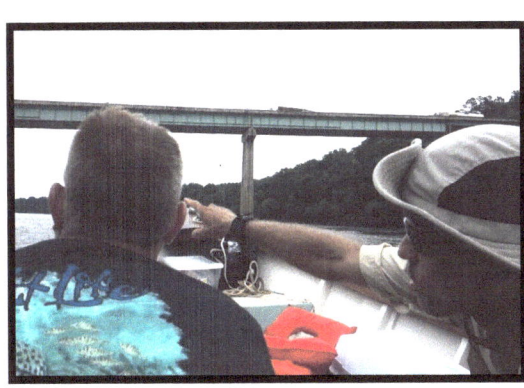

Figure 44: Scenes from the 2015-16 Contact Archeology Field Season.
Contact Archeology Inc. Files.

Figure 45: Scenes from the 2015-16 Contact Archeology Inc. Field Season.
Contact Archeology Inc. Files.

Figure 46: Scenes from the 2015-16 Contact Archeology Inc. Field Season.
Contact Archeology Inc. Files.

Epilogue

The documents and artifacts left behind by the Soto (1540) and Luna (1559) expeditions associated with their treks through Alabama and Northwest Florida are immensely important to our understanding of the anthropology of the European and Native cultures of the 1500s.

By recovering Native and Spanish artifacts and associated archeological features such as structures, town layouts, burial areas, plazas, campsites, and refuse pits, we are led to a better understanding of the lifeways of both cultures. By coupling the archeology with the thousands of pages of historic Spanish writings relative to their journeys and the Native peoples they encountered, an unprecedented research opportunity is within our grasp.

By identifying Native sites and Spanish settlements of the 16th Century, many research questions can be answered:

A sample of some the research questions follow.

- What were the geographic ranges of the Chiefdoms of Mabila and Tascaluza in Southern Alabama and the Chiefdom of Ochuse in Northwest Florida?

- Are there undiscovered historic documents and maps in today's archives?

- Were Native diseases transmitted to the Spaniards as well?

- How did the Native cultures change after the expeditions of Soto and Luna left the region?

- Were the Native Chiefdoms of Southwest Alabama and Northwest Florida in decline before the Spaniards arrived or did the arrival of the Europeans bring down the Chiefdoms?

- What foods were the Spaniards of the Soto and Luna expeditions eating to survive during their time in Southwest Alabama?

- How many Spaniards actually died in the Battle of Mabila and the Settlement of the Sacred Cross at Nanipacana and the colony on Pensacola Bay?

- Are there more details of the Soto and Luna expeditions as yet undiscovered in historic documents?

There is an axiom. Archeological research brings about many answers but it also brings about many questions as well.

Contact Archeology Inc. is conducting research in Southwest Alabama and Northwest Florida looking for the answers … and the questions.

References and Samples of Related Works

Author/s	Date	Title
Arnold, J. Barto, III and Robert Weddle	1978	*The Nautical Archeology of Padre Island: The Spanish Shipwrecks of 1554.* Academic Press. New York.
Atchison, Robert B.	1987	*Archaeological Survey in the Lower Cahawba Drainage.* University of Alabama, Office of Archaeological Research, Report of Investigations 53. The University of Alabama.
	1988	*Archaeological Survey of Selected Areas of the Alabama River Drainage, Dallas County, Alabama.* Office of Archaeological Research. Report submitted to the Alabama Historical Commission.
Ball, T.H.	1978	*A Glance into the Great Southeast or Clarke County, Alabama and Its Surroundings from 1540 to 1877.* Reprinted by the Clarke County Historical Society from the 1882 original.
Bigelow, A.	1851	*Observations on Some Mounds on the Tensaw River.* American Journal of Science, Article XXI, Vol. 65.
Blake, Alan	1988	*A Proposed Route for the Hernando de Soto Expedition, Based on Physiography and Geology.* Alabama De Soto Commission Working Paper No. 6. University of Alabama.
Blair, Elliot H., Lorann S.A. Pendleton, and Peter Francis, Jr.	2009	*The Beads of St. Catherines Island.* American Museum of Natural History Anthropological Papers, No. 89.
Bourne, Edward Gaylord	1904a	*Narratives of the Career of Hernando de Soto 1.* A.S. Barnes and Company. New York.
	1904b	*Narratives of the Career of Hernando de Soto 2.* A.S. Barnes and Company. New York.
Bowen Map	1764	*A New Map of Georgia with Part of Carolina, Florida, and Louisiana.* Copy in the Agee map collection, Birmingham Public Library.
Brame, James Y.	1928	*De Soto in Alabama.* Arrow Points 13. Montgomery, Alabama.
Brannon, Peter A.	1921	*The Route of De Soto from Cosa to Mauvilla.* Arrow Points 2, No. 1. Montgomery, Alabama.
Brown, Ian W.	2003	*Bottle Creek: A Pensacola Culture Site in South Alabama.* University of Alabama Press.
Brown, Ian W. and Richard S. Fuller (eds.)	1993	*Bottle Creek Working Papers on the Bottle Creek Site (1Ba2), Baldwin County, Alabama.* Journal of Alabama Archaeology 39, Nos. 1-2.

Brain, Jeffrey P.   1975   *Artifacts of the Adelantado*. In, *Conference on Historic Site Archaeology Papers 1973.*

1985a   *The Archeology of the De Soto Expedition*. In, *Alabama and the Borderlands: From Prehistory to Statehood.* (eds.) R. Badger and L. Clayton. The University of Alabama Press.

1985b   *Introduction: Update of De Soto Studies Since the United States De Soto Expedition Commission Report*. In, *Final Report of the United States De Soto Expedition Commission.* Smithsonian Institution Press.

Brown, Ian W.   2002   *An Archaeological Survey in Clarke County*. Submitted to the Alabama Historical Commission by The Gulf Coast Survey, Alabama Museum of Natural History, University of Alabama.

2003a   (ed.) *Bottle Creek: A Pensacola Culture Site in South Alabama.* The University of Alabama Press. Tuscaloosa.

Brown, Ian W. and   1993   *A Preliminary Report on Gulf Coast Survey Excavations at the Bottle*
Richard S. Fuller   *Creek Site, 1991.* Journal of Alabama Archaeology 39(1-2): 151-169.

Burke, R.P.   1936   *Check List of Indian Glass Trade Beads*. Arrow Points 21. Montgomery, Alabama.

Cabeza de Vaca, Alvar   1973   *The Journey of Alvar Nunez Cabeza de Vaca and His Companions*
Nunez   *from Florida to the Pacific, 1528-1536.* Translated by Fanny Bandelier. AMS Press. New York. First AMS edition in 1973 from the original Allerton Book Co. edition of 1922. New York.

Chardon, Roland   1980   *The Elusive Spanish League: A Problem of Measurement in the Sixteenth-Century New Spain.* The Hispanic American Historical Review 60 (2): 294-302.

Chase, David   1983   *Site 1Ds53: A Glimpse of Central Alabama Prehistory from the Archaic Period to the Historic Period*. In, Archaeology in Southwestern Alabama. Edited by Caleb Curren. Alabama-Tombigbee Regional Commission. Camden, Alabama.

Clayton, Lawrence A.,   1993   *The De Soto Chronicles: The Expedition of Hernando de Soto to North*
Vernon James Knight,   *America in 1539-1543* (2 Vols.). The University of Alabama Press.
Jr., and Edward C.
Moore (editors)

Cottier, John W.   1968   *Archaeological Salvage Investigations in the Millers Ferry Lock and Dam Reservoir.* Report to the U.S. Department of the Interior, National Park Service from the University of Alabama.

1970   *The Alabama River Phase: A Brief Description of a Late Phase in the Prehistory of South Central Alabama.* On file at the Department of Anthropology, The University of Alabama.

Cottier, John W. and   1980   *Interim Report of an Archaeological Survey of U.S. Army Corps of*
Craig T. Sheldon   *Engineers Properties along the Alabama River.* Report to the U.S. Army Corps of Engineers Mobile District from Auburn University.

Crenay, Baron de   1733   *Carte de Partie de la Louisianne qui Comprend le Cours du Missipy.*

Crosby, Alfred W. Jr.   1972   *The Columbian Exchange: Biological and Cultural Consequences of 1492.* Greenwood Press. Westport, Connecticut.

Cumming, William P.   1958   *The Southeast in Early Maps.* Princeton University Press. Princeton, New Jersey.

Curren, Caleb   1982   *The Alabama River Phase: A Review.* In, Archaeology in Southwest Alabama: A Collection of Papers. Alabama-Tombigbee Regional Commission. Camden, Alabama.

                            1984   *The Protohistoric Period in Central Alabama.* Alabama-Tombigbee Regional Commission. Camden, Alabama.

                            1986a   *An Archaeological Reconnaissance of Choctaw, Washington, and Southern Clarke Counties in Southwest Alabama.* Report to the Alabama Historical Commission from the Alabama-Tombigbee Regional Commission. Camden.

                            1986b   *In Search of De Soto's Trail. Alabama-Tombigbee Regional Commission. Bulletins of Discovery 1.* Camden.

                            1987   *The Route of the Soto Army Through Alabama.* Alabama De Soto Commission Working Paper Series 3. Tuscaloosa.

                            1988   *A Rebuttal of the "Georgia Reconstruction" of the Soto Route Through Alabama.* Report submitted to the Alabama De Soto Commission. Tuscaloosa.

                            1991a   *Spades Are Trumps.* The Soto States Anthropologist 91(1):39-44. Tallahassee.

                            1991b   *An Attempt to Legislate History: The De Soto Commission and the National Park Service in Action.* The Soto States Anthropologist 91 (2):126-148. Tallahassee.

                            1992   *Archeology in the Mauvila Chiefdom, Native and Spanish Contacts during the Soto and Luna Expeditions.* Mobile Historic Development Commission. Mobile.

                            2007   *Going South Towards the Sea: A Radical Look at Soto's Route through South Alabama.* Archeology Ink: An Online Research Journal. (archeologyink.com).

                            2011   *Archeological Excavations at an Early Mississippian Mound in Central Alabama.* Archeology Ink: An Online Research Journal. (archeologyink.com).

2012   *An Exploration for the Oldest European Colony on the American Gulf Coast (AD 1559).* Archeology Ink: An Online Research Journal. (archeologyink.com).

2013   *Artifactual Remains of the Spanish Conquistadors: Alabama and Northwest Florida.* Archeology Ink: An Online Research Journal. (archeologyink.com).

2014   *A Campsite of Tristan de Luna on Mobile Bay?* Archeology Ink: An Online Research Journal. (archeologyink.com).

2016a   *The Discovery of The 1559 Luna Colony in Pensacola, Florida: the Evidence?* Archeology Ink: An Online Research Journal. (archeologyink.com).

2016b   *A Cartographic Template for the 1559 Spanish Luna Colony?* Archeology Ink: An Online Research Journal. (archeologyink.com).

2016c   *Anchorage or Grounding? Two Shipwrecks of the 1559 Luna Expedition, Pensacola Bay, Florida.* Archeology Ink: An Online Research Journal. (archeologyink.com).

Curren, Caleb and Mark Curl

1985   *An Archaeological Reconnaissance of Wilcox and Dallas Counties in Southwest Alabama.* Report submitted to the Alabama Historical Commission from the Alabama-Tombigbee Regional Commission.

Curren, Caleb, George Lankford and Gregory Spies

1971   *Progress Reports 1-2 of an Archaeological Site Survey of South Alabama.* Report to the Archaeological Research Association of Alabama, Inc. from the University of South Alabama.

Curren, Caleb, Keith J. Little and George E. Lankford, III

1981   *The Route of the Expedition of Hernando de Soto Through Alabama.* Paper presented at the 38th Annual Meeting of the Southeastern Archaeological Conference, Nov. 12-14, 1981. Asheville, North Carolina.

1982   *Archaeological Research Concerning Sixteenth Century Spanish and Indians in Alabama.* Report on file at the Alabama-Tombigbee Regional Commission.

Curren, Caleb and Janet Lloyd

1987   *Archeological Survey in Southwest Alabama: 1984-1987.* Alabama-Tombigbee Regional Commission Technical Report 1.

Curren, Caleb and Lee McKenzie

1988   *Archaeological Investigations at Three Sites in the "Mauvila Province."* Report to the Alabama Historical Commission from the Alabama-Tombigbee Commission.

Curren, Caleb and Rhonda Majors

1984   *An Archaeological Reconnaissance of Monroe and Clarke Counties in Southwest Alabama.* Report submitted to the Alabama Historical Commission from Alabama-Tombigbee Regional Commission.

Curren, Caleb and Noel R. Stowe

1974   *Progress Reports 1-2 of an Archaeological Site Survey of South Alabama.* Report to the Archaeological Research Association of Alabama Inc. from the University of South Alabama.

Curren, Caleb, Keith J. Little and Harry 0. Holstein

1989   *Aboriginal Societies Encountered by the Tristan de Luna Expedition.* The Florida Anthropologist 42(4):38 1-395.

Curren, Caleb, Steve Newby, and Erick Mosley

2001   *Archeological Excavations at an Early Mississippian Mound (1Ds172) in Central Alabama.* archeologyink.com (An Online Archeological Journal).

Deagan, Kathleen A.

1979   *The Material Assemblage of 16th-Century Spanish Florida.* Historical Archaeology 12:25-50.

1980   *Spanish St. Augustine: America's First "Melting Pot."* Archaeology 33(5):22-30.

1981   *Downtown Survey: The Discovery of Sixteenth Century St. Augustine in an Urban Area.* American Antiquity 46(3):626-634.

1987   *Artifacts of the Spanish Colonies of Florida and the Caribbean, 1500-1800, Vol. 1: Ceramics, Glassware, and Beads.* Smithsonian Institution Press.

Deagan, Kathleen and David Hurst Thomas (eds. / contributors)

2009   *From Santa Elena to St. Augustine: Indigenous Ceramic Variability (A.D. 1400-1700).* Anthropological Papers of the American Museum of Natural History No. 90.

DeJarnette, David L. (ed.)

1958   *An Archaeological Study of a Site Suggested as the Location of the Upper Creek Indian Community of Coosa Visited by Hernando de Soto in 1540.* Masters Thesis, University of Alabama.

1975   *Highway Salvage Excavations at two French Colonial Period Indian Sites on Mobile Bay. Alabama.* Report to the Alabama Highway Department from The University of Alabama.

DeLeon, Mark F.

1976   *Archaeological Investigations in the Rother L. Harris Reservoir: 1976.* Report to the Alabama Power Company from The University of Alabama.

De L'isle, G.

1701   *Carte Des Environs Du Mississippi. Carte de la Louisiane.*

1718   *Carte de la Louisiane.*

DePratter, Chester, Charles Hudson and Marvin Smith   1985   *The Hernando de Soto Expedition: From Chiaha to Mabila.* In, Alabama and Its Borderlands, from Prehistory to Statehood. R. Badger and L.A. Clayton editors. University of Alabama Press. University, Alabama.

De Vorsey, Louis Jr.   1971   *Early Maps as a Source in the Reconstruction of Southern Indian Landscapes. In, "Red, White, and Black."* Charles M. Hudson (ed.). Southern Anthropological Proceedings 5.

Dickens, Roy S. Jr.   1971   *Archaeology in the Jones Bluff Reservoir of Central Alabama.* Journal of Alabama Archaeology 17(1):3-102.

Dobyns, Henry F.   1983   *Their Number Became Thinned: Native American Population Dynamics in Eastern North America.* University of Tennessee Press, Knoxville, TN.

Eubanks, W.S. Jr.   1990   *Swanton: Four.... Hudson: Zero.* The Soto States Anthropologist 90(1):3-32.

1991a   *Artifacts and Spanish Explorers.* The Soto States Anthropologist 91(1):86-92.

1991b   *De Soto, Still Another Look.* The Soto States Anthropologist 91(2):149-190.

1991c   *Adventures in Spanish and Spanish Adventurers.* The Soto States Anthropologist 9 1(3):228-236.

1991d   *Legions of Leagues.* The Soto States Anthropologist 91(4):299-333. 1992

Fairbanks, Charles H   1955   *An Investigation of Prehistoric Processes on the Gulf Coastal Plain: Final Report.* A Report to the National Science Foundation.

Finlay, Louis M. Jr.   1991   *The Syphrit Coin.* Clarke County Historical Society Quarterly 16(2):8-12.

Fuller, Richard S.   1985   *The Bear Point Phase of the Pensacola Variant: The Protohistoric Period in Southwest Alabama.* The Florida Anthropologist 38:1-2.

Fuller, Richard S. and Noel R. Stowe   1982   *A Proposed Typology for Late Shell Tempered Ceramics in the Mobile Bay / Mobile-Tensaw Delta Region.* In, Archeology in Southwest Alabama: A Collection of Papers. Caleb Curren (ed.). Alabama-Tombigbee Regional Commission. Camden, Alabama.

Fuller, Richard S., Diane E. Silvia, and N.R. Stowe   1984   *The Forks Project: An Investigation of the Late Prehistoric-Early Historic Transition in the Alabama-Tombigbee Confluence Basin.* Report submitted to the Alabama Historical Commission from The University of Alabama.

Graham, J. Bennett   1967   *A Preliminary Report of Salvage Archaeology in the Claiborne Dam Reservoir.* Report to the National Park Service from the University of Alabama Department of Anthropology.

Gresham, Thomas H., Karen Wood, Jerald Ladbetter, Chad Braley, and Paul Gradner	1987	*Archaeological Testing at 1Mn30 Eureka Landing, Monroe County, Alabama.* Report from Southeastern Archaeological Services to U.S. Army Corps of Engineers.
Hally, David J.	1993	*The Territorial Size of Mississippian Chiefdoms.* Archaeology of Eastern North America: Papers in Honor of Stephen Williams. Archaeological Report No. 25. Mississippi Department of Archives and History.
Hamilton, Peter J.	1976	*Colonial Mobile.* Reprinted from the 1910 original. University of Alabama Press. University, Alabama.
Harrisse, Henry	1892	*The Discovery of North America.* London and Paris.
Hastings, Wink (ed.)	1990	*A Synopsis of the Hernando De Soto Expedition, 1539-1545. De Soto Trail National Historic Trail Study. Final Report.* National Park Service Southeast Region. Atlanta.
Hayward, Hampton Dart, Caleb Curren, Ned J. Jenkins and Keith J. Little	1990	*Archeological Investigations in the Proposed Pafallaya Province.* Report Submitted to the Alabama De Soto Commission.
Hawkins, Benjamin	1938	*The Creek Country.* Americus Book Company. Americus, Georgia.
Higginbotham, Jay	1966	*The Mobile Indians.* Mobile, Alabama.
	1977	*Old Mobile: Fort Louis de la Louisiana, 1702-1711.* Museum of the City of Mobile. Mobile, Alabama.
Hill, Mary C.	1979	*The Alabama River phase: A Biological Synthesis and Interpretation.* M.A. thesis, Department of Anthropology, University of Tennessee. Knoxville.
	1981	*The Mississippian Decline in Alabama.* Paper presented at the 50th Annual Meeting of the American Association of Physical Anthropologists, Detroit, Michigan. Paper Abstracted in American Journal of Physical Anthropology 54:233.
	1996	*Protohistoric Aborigines in West-Central Alabama: Probable Correlations to Early European Contact.* In, Bioarchaeology of Native American Adaptation in the Spanish Borderlands. (eds.) B.F. Baker and L. Kealhofer. Pgs. 17-37. University Press of Florida. Gainesville.
	2001	*Porotic Hyperostosis as an Indicator of Anemia: An Overview of Correlation and Cause.* PhD. Dissertation. University of Massachusetts. Amherst.
Hill, Mary C. and G.A. Clark	1981	*Skeletal Analysis of Burials from 1Tu4 and 1Ha19.* In, *Archaeological Investigations (1933-1980) of the Protohistoric Period of Central Alabama.* (by) Caleb Curren and Keith Little. Alabama-Tombigbee Regional Commission. Camden, Alabama. (pgs.) 135-184, 200-233, 281-296.

	1984	*Biological Burial Analysis.* In, *The Protohistoric Period in Central Alabama.* Alabama-Tombigbee Regional Commission. Camden.
Holmes, Nicholas H. Sr.	1993	*The De Luna Expedition.* The Soto States Anthropologist 93(3-4): 95-111.
Holmes, Nicholas H. Jr.	1963	*The Site on Bottle Creek.* Journal of Alabama Archaeology 9(1). Tuscaloosa.
	1993	*A Geometric Test of Some Proposed De Soto Routes Through Alabama.* The Soto States Anthropologist 93(3-4): 114-121.
Holmes, Nicholas H. Jr. and Charles E. Bates	1996	*A Comparison of Trace Elements Present in Two Iron Objects from a Mississippian Site in South Alabama, with Those of Spanish, French, and American Irons.* Journal of Alabama Archaeology 42(2):154-171.
Holmes, Nicholas H. III, Douglas Jones, and Alan Blake	1992	*Pigmullion.* Soto States Anthropologist 92(3-4).
Holmes, Andrew	2004	*A Critical Analysis of Past Inquires into the Location of the Battle Site of Mauvilla, A Comprehensive Review of the De Soto Chronicles Pertaining to the Nature and Location of Mauvilla, and My Conclusions on the Location of Mauvilla and a Description of What has Been Found There.* Manuscript on file.
Hudson, Charles, Marvin T. Smith, Chester B. DePratter, and Emilia Kelley	1989	*The Tristan de Luna Expedition, 1559-1561.* Southeastern Archaeology 81:31-45.
Hudson, Charles	1997	*Knights of Spain, Warriors of the Sun: Hernando de Soto and South's Ancient Chiefdoms.* University of Georgia Press. Athens.
	1988	*Critique of Little and Curren's Reconstruction of De Soto's Route through Alabama.* Alabama De Soto Commission Working Paper No. 12. Tuscaloosa.
	1989	*De Soto in Alabama.* Alabama De Soto Commission Working Paper No. 10.
Hudson, Charles, Marvin T. Smith, and Chester DePratter	1990	*The Hernando de Soto Expedition: From Mabila to the Mississippi River.* In, *Towns and Temples along the Mississippi.* (eds.) David H. Dye and Cheryl Anne Cox, pp. 181-207. The University of Alabama Press.
Jenkins, Ned J. and Teresa Paglione	1980	*An Archaeological Reconnaissance of the Lower Alabama River.* Report to the Alabama Historical Commission from Auburn Univ.
	1983	*Lower Alabama River Ceramic Chronology: A Tentative Assessment.* In, Archaeology in Southwestern Alabama: A Collection of Papers. Edited by Caleb Curren. Alabama-Tombigbee Regional Commission. Camden, Alabama.
	2008	*Mabila Reconnaissance, 2007.* Report to the University of Alabama and the Alabama Historical Commission.

Jenkins, Ned J.	2014	*The Hernando de Soto and Tristan de Luna Expeditions in Central Alabama, 1540-1560: Routes, Cultures and Consequences.* Manuscript on file, Auburn University.
Jeter, Marvin D.	1973	*An Archaeological Survey in the Area East of Selma, Alabama, 1971-1972.* Report to the Alabama Historical Commission from the University of Alabama Birmingham.
Johnson, Adrian	1974	*America Explored.* The Viking Press.
Knight, Vernon James Jr. (ed.)	2009	*The Search for Mabila.* University of Alabama Press.
Knight, Vernon James Jr., and Sheree L. Adams	1981	*A Voyage to the Mobile and Tomeh in 1700, with Notes on the Interior of Alabama.* Ethnohistory 28(2): 179-194.
Knight, Vernon James, Lucinda Freeman, Craig T. Sheldon Jr., Kreen Hawsey, and Gregory A. Waselkov	2015	*White Oak Creek Archaeology in Dallas County, on the Trail of DeSoto.* Report to the Alabama Power Company Foundation on Grant 2014-48821 from the University of Alabama.
Lankford, George E. III	1977	*A New Look at De Soto's Route through Alabama.* Journal of Alabama Archaeology 23(1): 10-36.
	1983	*A Documentary Study of Native American Life in the Lower Tombigbee Valley. In, Cultural Resources Reconnaissance Study of the Black Warrior-Tombigbee System Corridor Alabama. Vol. II, Ethnohistory.* Report to the U.S. Army Corps of Engineers Mobile District from the University of South Alabama.
Little, Keith J.	1988	*The De Soto Route Debates: A Presentation for the General Public.* Bulletins of Discovery No. 2. Alabama –Tombigbee Regional Commission. Camden, Alabama.
	2008	*European Artifact Chronology and Impacts of Spanish Contact in the Coosa River Valley.* PhD. Dissertation, University of Alabama.
Little, Keith J. and Kevin Harrelson	2005	*Pine Log Creek: Ethnohistoric Archaeology in the Alabama-Tombigbee Confluence Basin.* Jacksonville State University Archaeological Laboratory Research Series No. 3.
Little, Keith J. and Caleb Curren	1990	*Conquest Archaeology of Alabama.* In Archaeological and Historical Perspectives on the Spanish Borderlands East, Columbian Consequences 2, David Hurst Thomas (ed). Smithsonian Institution Press.
Lowery, Woodbury	1959	*The Spanish Settlements Within the Present Limits of the United States.* 1513-1561. Russell and Russell, Inc. New York, New York.
Martin, Troy O.	1989	*Archaeological Investigations of an Aboriginal Defensive Ditch at Site 1Ds32.* Journal of Alabama Archaeology, 35:1.
McWilliams, Richebourg G. (ed.)	1981	*Iberville's Gulf Journals.* University of Alabama Press. University, Alabama.

Michaelis, Ronald F.   1978   *Old Domestic Base-Metal Candlesticks.* Antique Collectors' Club, Woodbridge, Suffolk, U.K.

Mikell, Gregory A. and Keith J. Little   1999   *An Archaeological Investigation of the Doctor Lake Site (1Ck219): A Search for the Sixteenth-Century Mabila Battlefield.* Journal of Alabama Archaeology 45 (1): 1-36.

Milanich, Jerald T.   1990   *The European Entrada into La Florida: An Overview.* In, *Columbian Consequences.* (ed.) David Hurst Thomas.

Mitchem, Jeffrey M. and B.G. McEwan   1988   *New Data on Early Bells from Florida.* Southeastern Archaeology 7 (1)39-48.

Moore, Clarence B.   1899   *Certain Aboriginal Remains of the Alabama River.* Journal of the Academy of Natural Sciences of Philadelphia 11 (4):289-347.

1901   *Certain Aboriginal Remains of the Tombigbee River.* Journal of the Academy of Natural Sciences of Philadelphia 13.

1904   *Aboriginal Urn-burial in the United States.* American Anthropologist 6:660-669.

1905a   *Certain Aboriginal Remains of the Black Warrior River.* Journal of the Academy of Natural Sciences of Philadelphia 13 (2):145-244.

1905b   *Certain Aboriginal Remains of the Lower Tombigbee River.* Journal of the Academy of Natural Sciences of Philadelphia 13(2):245-276.

Nance, D. Roger   1976   *The Archaeological Sequence at Durant Bend, Dallas County, Alabama.* Alabama Anthropological Society, Special Publication 2. Orange Beach.

Nielsen, Jerry J., John W. O'Hear and Charles W. Moorehead   1973   *An Archaeological Survey of Hale and Green Counties, Alabama.* Report to the Alabama Historical Commission from The University of Alabama.

Nielsen, Jerry J.   1969   *Archaeological Investigations of Three Additional Sites in the Claiborne Lock and Dam Reservoir.* Report submitted to the National Park Service from the University of Alabama.

Oakley, Carey B. and G. Michael Watson   1977   *Cultural Resources Inventory of the Jones Bluff Lake, Alabama River, Alabama.* Report of Investigations 4, Office of Archaeological Research. The University of Alabama.

Owen, Thomas M.   1921   *History of Alabama Dictionary of Alabama Biography Vol. II.* S.J. Clarke Publishing Co. Chicago, Illinois.

Padilla, Fray Agustin Davila   1596   *Historia de la Fundacion y Discorso de la Provincia de Santiago de Mexico de la Orden de Predicadores.* Madrid.

Palmer, Edward   1884   Unpublished field notes on file at the Alabama Archives and History. Montgomery, Alabama.

Payne-Gallwey, Ralph   1995   *The Book of the Crossbow.* Dover Press.

Pickett, Albert J.   1851   *History of Alabama.* Walker and James Charlestown.

Priestley, Herbert I.   1928   *The Luna Papers.* Books for Libraries Press. Freeport, New Jersey.

                                 1936   *Tristan de Luna. Conquistador of the Old South.* Arthur H. Clarke Company. Glendale, California.

Sears, William H.   1959   *An Investigation of Prehistoric Processes on the Gulf Coastal Plain.* Report to the National Science Foundation from the University of Florida.

                                 1977   *Prehistoric Culture Areas and Culture Change on the Gulf Coastal Plain.* In, Research Essays in Honor of James B. Griffin, Museum of Anthropology. University of Michigan. Anthropological Papers 61.

Sheldon, Craig T.   1974   *The Mississippian-Historic Transition in Central Alabama.* Ph.D. Dissertation. University of Oregon.

                                 2001   *The Southern and Central Alabama Expeditions of Clarence Bloomfield Moore.* University of Alabama Press.

Sheldon, Craig T. and Ned J. Jenkins   2014   *Artifacts of the de Soto and de Luna y Arellano Expeditions in Alabama.* Manuscript on file, Auburn University.

Smith, Hale   1956   *The European and the Indian: European-Indian Contacts in Georgia and Florida.* Florida Anthropological Society Publications 4:13-15. Gainesville, Florida.

Smith, Marvin T.   1976   *The Route of DeSoto Through Tennessee, Georgia, and Alabama: The Evidence from Material Culture.* Early Georgia 4(1-2):27-48.

                                 1982   *Chronology from Glass Beads: The Spanish Period in the Southeast.* Paper presented at the Glass Trade Bead Conference, Rochester Museum and Science Center.

                                 1984   Depopulation and Culture Change in the Early Historic Period of the Interior Southeast. Ph.D. Dissertation. University of Florida.

                                 1987   *Archaeology of Aboriginal Culture Change in the Interior Southeast: Depopulation during the Early Historic Period.* Ripley P. Bullen. Monographs in Anthropology and History 6. Florida State Museum. Gainesville.

Smith Marvin T. and Mary Elizabeth Goode   1982   *Early Sixteenth Century Glass Beads in the Spanish Colonial Trade.* Cottonlandia Museum. Greenwood, Mississippi.

Smith, Roger C., John R. Bratten, J. Cossi, and Keith Plaskett.   1998   *The Emanuel Point Ship Archaeological Investigations, 1997-1998.* Report of Investigations 68. Archaeology Institute, University of West Florida.

Stowe, Noel R.   1971   University of South Alabama - Archaeological Research Association of Alabama. Inc. Site Survey of South Alabama. Report on file with the Archaeological Research Association of Alabama, Inc.

                                1978   *A Preliminary Cultural Resources Literature Search of the Mobile Tensaw Bottomlands.* Report on file, National Park Service, Tallahassee, Florida.

                                1985   *The Pensacola Variant and the Bottle Creek Phase.* The Florida Anthropologist 38(1-2).

South, Stanley   1980   *The Discovery of Santa Elena.* Institute of Archeology and Anthropology, University of South Carolina. Research Manuscript Series 165. Columbia, South Carolina.

                                1981   *Exploring Santa Elena.* Institute of Archeology and Anthropology. University of South Carolina. Research Manuscript Series 184. Columbia, South Carolina.

                                1983   *Revealing Santa Elena.* Institute of Archeology and Anthropology. University of South Carolina. Research Manuscript Series 188. Columbia, South Carolina.

                                1984a   *Testing archaeological Sampling Methods at Fort San Felipe.* Institute of Archeology and Anthropology. University of South Carolina, Research Manuscript Series 190. Columbia, South Carolina.

                                1984b   *Excavation of the Casa Fuerte and Wells at Ft. San Felipe 1984.* Institute of Archeology and Anthropology, University of South Carolina, Research Manuscript Series 196. Columbia, South Carolina.

                                1988   *Spanish Artifacts from Santa Elena.* Anthropological Studies 7. Occasional Papers of the South Carolina Institute of Archaeology and Anthropology. University of South Carolina.

Stowe, Noel R., Richard   1982   *A Preliminary Report on the Pine Log Creek Site (1Ba462).* Report on
Fuller, Amy Snow, and   file, University of South Alabama Archaeology Lab. Mobile, Alabama.
Jennie Trimble

Swanton, John R.   1922   *Early History of the Creek Indians and Their Neighbors.* Smithsonian Institution Bureau of American Ethnology Bulletin 73. Indians of the Southeastern United States. Bureau of American Ethnology Bulletin 137.

Thomas, Cyrus   1894   *Report of the Mound Explorations of the Bureau of Ethnology.* Smithsonian Institution. Bureau of Ethnology Annual Report 1890-1981(12).

Thomas, David Hurst   1988   *St. Catherines: An Island in Time.* Georgia History and Culture Series. Atlanta: Georgia Endowment for the Humanities.

                                1993   *Historic Indian Period Archaeology of the Georgia Coastal Zone.* Georgia Archaeological Research Design Paper 8.

                                1990   *Columbian Consequences.* (ed.). Smithsonian Institution Press.

Trickey, E. Bruce   1958   *A Chronological Framework for the Mobile Bay Region.* American Antiquity 23(4).

                             1995   *Mauvilla: A New Approach.* Journal of Alabama Archaeology 41: 79-87.

Trickey, E. Bruce   1971   *A Chronological Framework for the Mobile Bay Region.* Journal of and Nicholas H. Holmes,   Alabama Archaeology 17(2).
Jr.

Varner, John and Jean-   1951   *The Florida of the Inca.* University of Texas Press. Austin, Texas.
nette Varner

Willey, Gordon R.   1949   *Archaeology of the Florida Gulf Coast.* Smithsonian Miscellaneous Collection 113. Washington D.C.

Wilson, Eugene M.,   2009   *Tracing De Soto's Trail to Mabila.* In, *The Search for Mabila.* (ed.)
Douglas E. Jones, and   Vernon James Knight Jr. University of Alabama Press.
Neal G. Lineback

Wimberly, Stephen B.   1960   *Indian Pottery from Clarke County and Mobile County, Southern Alabama.* Alabama Museum of Natural History Museum Paper 36. University, Alabama.

www.ingramcontent.com/pod-product-compliance
Lightning Source LLC
Chambersburg PA
CBHW050739180526
45159CB00003B/1282